MANAGING THE EARTH

Managing the Earth

The Linacre Lectures 2001

Edited by
JAMES C. BRIDEN
and
THOMAS E. DOWNING

OXFORD
UNIVERSITY PRESS

OXFORD
UNIVERSITY PRESS

Great Clarendon Street, Oxford OX2 6DP

Oxford University Press is a department of the University of Oxford.
It furthers the University's objective of excellence in research, scholarship,
and education by publishing worldwide in

Oxford New York

Auckland Bangkok Buenos Aires Cape Town Chennai
Dar es Salaam Delhi Hong Kong Istanbul Karachi Kolkata
Kuala Lumpur Madrid Melbourne Mexico City Mumbai Nairobi
São Paulo Shanghai Singapore Taipei Tokyo Toronto

with an associated company in Berlin

Oxford is a registered trade mark of Oxford University Press
in the UK and in certain other countries

Published in the United States
by Oxford University Press Inc., New York

© Oxford University Press, 2002 with the exception of:
Chapter 2 *Risks of Conflict* © Crispin Tickell 2002

The moral rights of the author have been asserted
Database right Oxford University Press (maker)

First published 2002

British Library Cataloguing in Publication Data
Data available

Library of Congress Cataloging in Publication Data
Managing the earth: the Linacre Lectures 2001/edited by
James C. Briden and Thomas E. Downing.
p. cm.
1. Environmental management. 2. Nature—Effect of human beings on. 3. Conservation
of natural resources. I. Briden, James C. II. Downing, Thomas E.
GE300.M362 2002 33.7'2—dc21 2002067160
ISBN 0-19-925267-X

1 3 5 7 9 10 8 6 4 2

Typeset in Stempel Garamond
by Hope Services (Abingdon) Ltd.
Printed in Great Britain
on acid-free paper by
Biddles Ltd.,
Guildford and King's Lynn

ACKNOWLEDGEMENTS

The eleventh annual series of Linacre Lectures was organized and sponsored by Linacre College, Oxford, and the Environmental Change Institute, University of Oxford. The lectures were presented in the Zoology Department of the University in January to March 2001. Our thanks are due to the Principal of Linacre College, Professor Paul Slack, and the Governing Body for their help and support.

J.C.B. and T.E.D.

CONTENTS

NOTES ON THE CONTRIBUTORS

James C. Briden is Professor of Environmental Studies and Director of the Environmental Change Institute, University of Oxford, and a Professorial Fellow of Linacre College, Oxford. He is author of over ninety papers on geophysics and palaeoclimatology and has been co-author of two books on the past configuration of the continents.

Lord (John) Browne is the Chief Executive Officer of BP. He is a graduate of Cambridge and Stanford Business School, a non-executive Director of Intel and Goldman Sachs, and a Trustee of the British Museum.

Robert Costanza is the Gund Professor of Ecological Economics at the University of Vermont's School of Natural Resources and Director of UVM's Gund Institute for Ecological Economics. He was previously Director of the University of Maryland Institute for Ecological Economics, and a Professor in the Center for Environmental Science. He is co-founder and past President of the International Society for Ecological Economics (ISEE), and is currently President of the International Society for Ecosystem Health. He is the author or co-author of over 300 scientific papers and sixteen books.

Thomas E. Downing is Senior Research Fellow of Linacre College. Until 2002 he was Reader in Climate Policy in the Environmental Change Institute, University of Oxford, and is now Director of the Oxford Office of the Stockholm Environment Institute. His major interests are vulnerability and adaptation to climate change and climatic hazards, using methods in participatory integrated assessment. He has published over 100 papers, books, reports, and book reviews.

Joyeeta Gupta is presently co-interim head of the Department of Environment Policy Analysis at the Institute for Environmental Studies in Amsterdam and leads the programme on International Environmental Agreements. She has written and co-edited six books on climate change and related fields.

Hermann Held is project leader at the Potsdam Institute for Climate Impact Research (PIK) on non-linear effects and uncertainty propagation in the Earth System, also Scientific Officer for IGBP-GAIM

(International Geosphere Biosphere Programme—Global Analysis, Integration and Modelling). He is a former Fellow of the Max Planck Society and the Alexander von Humboldt Foundation on non-linear effects in complex systems.

Bert Metz is head of the International Environmental Assessment Division at RIVM, the Netherlands National Institute for Public Health and Environment, and co-chairman of the Working Group on Mitigation of the United Nations Intergovernmental Panel on Climate Change. He was extensively involved in the Climate Change Convention and the Kyoto Protocol; he was responsible for the IPCC's Third Assessment Report on Mitigation and Special Reports on Technology Transfer, Emission Scenarios, and Aviation.

Philippe Sands is Professor of Laws at University College London and Global Professor of Law at New York University School of Law. He is author of *Principles of International Environmental Law* and *Bowett's Law of International Institutions* and has written extensively on numerous areas of international law, particularly in relation to administration of international justice, and protection of the environment.

Hans-Joachim Schellnhuber is Professor for Theoretical Physics at the University of Potsdam, Director of the Potsdam Institute (PIK), and Visiting Research Director of the Tyndall Centre for Climate Change Research based at the University of East Anglia, Norwich. He is author or co-author of more than 150 articles and books on complex systems theory, environmental analysis, and integrated assessment.

Sir Crispin Tickell is Chancellor of the University of Kent at Canterbury. He was formerly British Ambassador to the United Nations, and Warden of Green College, Oxford. He has written and spoken widely on environmental issues, in particular climate and biological diversity.

LIST OF FIGURES

LIST OF TABLES

ABBREVIATIONS

bcf/d	billion cubic feet (of gas) per day
CITES	Convention on International Trade in Endangered Species of Wild Fauna and Flora
EMIC	Earth System model of intermediate complexity
ENSO	El Niño/Southern Oscillation
FDI	foreign direct investment
GCM	general circulation model
GDP	gross domestic product
GNP	gross national product
Gt	gigatonne
GtC	gigatonnes of carbon
GtCeq	gigatonnes of carbon equivalent
HDI	Human Dimensions Index
ICJ	International Court of Justice
IGBP	International Geosphere–Biosphere Programme
IHDP	International Human Dimensions Programme on Global
IPCC	Intergovernmental Panel on Climate Change
ka	thousand years ago
mb/d	million barrels (of oil) per day
NAFTA	North American Free Trade Agreement
NSI	national systems of innovation
ODA	official development assistance
ppmv	parts per million (by volume)
t	tonne
TED	turtle excluder device
UNCED	United Nations Conference on Environment and Development
WCRP	World Climate Research Programme

Introduction

James C. Briden and Thomas E. Downing

SCARCELY a day goes by without news of drought or flood, famine or hurricane, conflict or fraught international negotiation to raise repeatedly the suspicion that humanity is living beyond the coping capacity of our host planet. The natural forces and human interests, individual and communal, are legion and apparently irreconcilable. How can we manage? How can we live sustainably on Earth, bequeathing to successor generations an environment at least as fit to live in as we currently enjoy? The issues go beyond the conventional bounds of politics and international diplomacy to the challenge of managing the Earth itself.

In some sectors, the challenge of managing the Earth has already been taken up. Pioneering in recent times was the Montreal Protocol on chlorofluorocarbons and the depletion of stratospheric ozone. More intractable has been proving that climate change is occurring and dangerous in the Intergovernmental Panel on Climate Change (IPCC), and then agreeing even the ineffective prescriptions of the Kyoto Protocol of the UN Framework Convention on Climate Change. Difficult as such sectoral issues are, and however successful present initiatives may be, they cannot be sufficient unless the interactions among the various sectors are addressed with the same degree of insight and comprehensive action.

The 2001 Linacre Lectures, organized and sponsored by Linacre College and the Environmental Change Institute in Oxford University, were conceived to throw light on the nature of these inter-sectoral and interdisciplinary challenges. Collectively, this book, based on the lectures, is intended to provide a perspective on the issue of managing the Earth that goes beyond the individual contributions. Our ambition is to provide a view of the prospects for achieving sustainable development, by defining the challenge in the broadest terms and then providing a selection of perspectives on the routes to solution. Challenges emerge that call for substantial leadership, research, and partnership if progress is to be made.

This volume is not a substitute for the detailed specialist assessments such as the IPCC. Rather, we seek to explore ground that may be less well covered in such assessments. For example, international science assessments may draw back from speculation that does not fit easily in the incremental, consensual paradigm. Or policy frameworks are seen as not science and left to other forums. Or issues of consumerism, equity, and business leadership are seen as value-laden and not subject to academic enquiry. Nor can this volume be comprehensive: issues of water resources and environmental health, for example, are addressed only in passing (though both are themes of other Linacre Lectures).

The book begins with the second Copernican revolution—examining the future of life on the Earth. Schellnhuber and Held (Chapter 1) note that the absence of life elsewhere in the universe (or our ability to detect it) might support the hypothesis that extinction is a real possibility. There is no solace in contemplating the likelihood that a self-sustaining Gaia is a necessary property of our planet. The first steps in observation and Earth Systems modelling have already been taken. Yet, the scientific challenges are daunting. Internationally coordinated research and action are imperative.

The present sustainability crisis is no less real. Tickell (Chapter 2) sees the disparity between rich and poor as one of the major driving forces of environmental degradation and a trigger for conflict. There are those who say that the events of 11 September 2001 derive in part from this root.

Costanza (Chapter 3) matches these visions of Earth's present and future with an exposition of the science needed to bridge 'objective' scientific analysis and values required for judgement and policy, and to study both ecosystems and economies. This science must be applied to world visions that reflect technological optimism or pessimism.

Technology is a bright prospect, but an uncertain salvation for climate change, according to Metz (Chapter 4). For technology to succeed it needs to be practical, socially acceptable, economical, and able to be marketed in current conditions. Significant barriers exist, including organizational capacity, the enabling environment of policy, and investment in long-term research.

A group of papers explores policy in additional detail. Browne (Chapter 5) provides a business view of technology. Innovation and corporate leadership are essential, but may not be socially acceptable on its own. Businesses are rarely very good at long-term thinking or strategic decision-making beyond a few years. They need an acceptable policy framework in which to operate. The backlash against globalization (and the World Trade Organization in particular) is a visible indicator of the need to rethink public–private partnerships.

Certainly the law is a necessary policy framework. Yet, Sands (Chapter 6) exposes the yawning gaps and inconsistencies in the international legal framework, which has simply not been designed to deal with the nuances that environmental conflicts frequently display. He discusses fascinating cases in which 'fairness' in terms of world trade confronts 'prudence' in conservation terms.

The achievement of fairness is elusive in international negotiations. Gupta (Chapter 7) reveals the disparities between parties in negotiating the Climate Change Convention. In a divided world international politics and diplomacy, as with law, are unsure but necessary foundations for managing the Earth.

The concluding paper applies lessons from a decade of food security research and applications to global change vulnerability. Downing asserts that often the major vulnerabilities are ignored—for instance, that imbalance between food self-sufficiency and international trade in a world where transport costs increase significantly. At the same time technical progress in vulnerability assessment, monitoring famine, and targeting food aid provides tools for preventing the worst impacts of climate change.

Finding solutions is clearly going to be desperately difficult. No single solution—whether a technology or international agreement—is going to be sufficient.

If the analysis of the Earth at risk, as set out at different scales by the authors of this volume, has merit, we need bold solutions that take us out of the technological, economic, and behavioural 'lock-in' that leads to a fragile and depleted Earth environment. Where do we start?

Perhaps some local experiments are worth global funding and might serve as the sort of learning experience that can be disseminated across the world. High-priority candidates would include fuel cell-based transport in India, community resilience and monitoring of vulnerability in semi-arid Africa, sustainable consumption in the United States.

Certainly partnerships are required. Are we satisfied with the efforts to link private and public interests? How might North–South partnerships develop to promote global sustainability?

The lack of organizational capacity and forums in which progress can be made is worrisome. What are the potential triggers to finding solutions? Will incremental developments of policy and economic instruments, like the emissions targets and greenhouse gas emissions trading in the Kyoto Protocol, be sufficient? What is the role of civil society?

A further force must come into play to a degree that has not been in evidence since 'civilization' began: that force is altruism. It is little addressed in this book, and rarely spoken of. It needs to emerge as a force at individual, communal, and international scales. If the role of academic discourse is to reveal gaps and failings as much as to display progress, that is surely a principal lesson to be drawn here.

We and the next two or three generations have the last chance to avert serious and irreversible downgrading of the habitability of Earth. Will we grasp the issues and act accordingly? In 2002 and the following years the eyes of the world will be on the second Earth Summit and on those responsible for acting upon it. By the time you read this you may already be able to deduce whether the Earth environment is moving towards sustainability or not, and whether future sustainability is joined to the present crisis in development.

I

How Fragile is the Earth System?

Hans-Joachim Schellnhuber and Hermann Held

THE SECOND COPERNICAN REVOLUTION

In July 2001 more than 1,800 scientists from 100 countries met in Amsterdam for the conference 'Challenges of a Changing Earth'. This event was something like an Earth Summit for Global Change science as it (i) brought together all pertinent parts and disciplines from the worldwide environmental research community, (ii) reviewed and summarized the state of the art in all fields that contribute to the growing understanding of the planetary machinery, and (iii) developed a programmatic vision—a scientific 'Agenda 21'—regarding future structure and advancement of research for global sustainability. The main messages and conclusions are compressed into the so-called Amsterdam Declaration, but a comprehensive and concise account is provided by the recent IGBP Science Paper entitled *Global Change and the Earth System* (Steffen and Tyson 2001). As one of us had the privilege to co-author this paper, we feel free to quote here a few crucial paragraphs:

The world faces significant environmental problems: shortages of clean and accessible freshwater, degradation of terrestrial and aquatic ecosystems, increases in soil erosion, loss of biodiversity, changes in the chemistry of the atmosphere, declines in fisheries, and the possibility of significant changes in climate. These changes are occurring over and above the stresses imposed by the natural variability of a dynamic planet and are intersecting with the effects of past and existing patterns of conflict, poverty, disease, and malnutrition.

The changes taking place are, in fact, changes in the human–nature relationship. They are recent, they are profound, and many are accelerating. They are cascading through the Earth's environment in ways that are difficult to understand and often impossible to predict. Surprises abound. At least, these human-driven changes to

The authors would like to thank W. von Bloh, C. Bounama, V. Brovkin, S. Franck, A. Ganopolski, T. Schneider von Deimling, and V. Petoukhov for helpful discussions.

the global environment will require societies to develop a multitude of creative response and adaptation strategies. Some are adapting already; most are not. At worst, they may drive the Earth itself into a different state that may be much less hospitable to humans and other forms of life.

The most important sentence in this citation is certainly the last one, which states that the planetary machinery may have a multitude of rather distinct modes of fundamental operation and that civilizatory interference with nature might inadvertently trigger transitions between these modes. For a number of reasons, such a statement would have been virtually unthinkable four or even three decades ago, when Earth-System-level analysis still appeared as mere science fiction. Today only few colleagues may be shocked by the following elaboration:

The interactions between environmental change and human societies have a long and complex history, spanning many millennia. They vary greatly through time and from place to place. Despite these spatial and temporal differences, in recent years a global perspective has begun to emerge that forms the framework for a growing body of research within the environmental sciences. Crucial to the emergence of this perspective has been the dawning awareness of two fundamental aspects of the nature of the planet. The first is that the Earth itself is a single system, within which the biosphere is an active, essential component. In terms of a sporting analogy, life is a player, not a spectator. Second, human activities are now so pervasive and profound in their consequences that they affect the Earth at a global scale in complex, interactive and accelerating ways; humans now have the capacity to alter the Earth System in ways that threaten the very processes and components, both biotic and abiotic, upon which humans depend.

Systems thinking and its application to the environment are not new. However, until very recently, much of the understanding about how the Earth operates was applied only to pieces (subcomponents) of the Earth. What is really new about the understanding of the Earth System over the last 10–15 years is a perspective that embraces the System *as a whole*. Several developments have led to this significant change in perception:

- The view of Earth from a spaceship, a blue-green sphere floating in blackness, triggers emotional *feelings* of a home teeming with life set in a lifeless void, as well as more analytical *perceptions* of a materially limited and self-contained entity.
- Global observation systems allow the application of *concepts* that were only previously applicable at subsystem level, or regional or local scales, to the Earth as a whole.
- Global databases allow global scale phenomena to be addressed with consistently acquired *data* that have the potential for harmonisation and comparison at a global scale.
- Dramatic advances in the power to infer characteristics of Earth System processes in the past allow contemporary observations to be viewed in a coherent *time continuum*.

- Enhanced computing power makes possible not only essential data assimilation, but increasingly sophisticated *models* improve understanding of functional interactions and system sensitivities.

Science has crossed the threshold of a profound shift in the perception of the human–environment relationship, operating across humanity as a whole and at the scale of the Earth as a single system.

So Earth System science seems to emerge as a transdisciplinary enterprise just in time to provide the cognitive basis for the sustainable development of human civilization. This emergence can be interpreted as a second Copernican revolution (Schellnhuber 1999)—both for epistemological and for socio-political reasons: the first Copernican revolution put the Earth in its correct astrophysical context, but it also shattered the medieval perception of our home planet as the hub of God's creation. As a consequence, the elites of the feudal hierarchy, who arrogated themselves central positions in this creation, lost their singularity in the eyes of the ordinary people, and the stage was set for the rise of the civil individual. The second Copernican revolution now discovers our planet as one whole and unique entity by watching it from outer space, by probing its digital caricatures in the virtual reality of cyberspace, or by reproducing parts of it in miniature. As a consequence, a deep consciousness of global interconnectedness is about to arise that reintegrates the individual into some planetary hierarchy—the *real* hierarchy of biogeochemical cycles, atmospheric circulation patterns, syndromes of land use change, international economic oscillations, and worldwide technological innovation waves. Thus, in a sense, the second revolution reverts and completes the first one.

And, if everything goes right, some 8 or 10 billion human beings will run this planet carefully and interactively and sustainably in the not too distant future. But, what if something goes terribly wrong? Is the Earth so fragile that it can be ruined by careless handling? And if so, what must we do, at the very least, to avoid this? Let us try to sketch a few tentative answers to these disturbing questions in the rest of this essay.

GAIA AND HER SISTERS

Before unfolding those answers, however, even more questions need to be asked: Is there anything special about the Earth System in the cosmic context? Do the universal laws of nature as generated in the Big Bang prescribe the emergence of bioplanets as the final stage of evolution? Or is Terra a mere whim of the Creation not entirely forbidden by its laws? The impulse of the original Copernican revolution is still with us and keeps on driving science and technology to explore the most distant pockets of the universe.

One of the fascinating opportunities arising from this development is the possibility of comparing our planet to a growing number of extra-solar ones (i.e. beyond the solar system).

Such a comparison can provide, in particular, a testbed for various hypotheses generated in the context of the famous and much debated GAIA theory (Lovelock 1995; Lovelock and Margulis 1974). This school of thought perceives the tightly coupled biotic and abiotic parts of the Earth as a kind of super-organism, which actively controls its own subsistence conditions and therefore keeps the planetary machinery in the biogeophysico-chemical regime appropriate for life. The validity of the Gaia approach clearly has a bearing on every scientific assessment regarding the probability and stability of 'Earth' in the sense of a galactic evolutionary concept.

One more practical aspect is the evaluation of complex global climate models, which have become indispensable tools for environmental policy advice yet are plagued with the obstinate problem of underdetermination: while 'hindcasting' of pertinent palaeo-records is, in principle, a very good option for validating a geophysical model, these records are rather sparse and unsystematic in view of the high-quality information needed. Therefore first principles of planetary operation beyond the fundamental laws of physics are badly needed, and Gaia heuristics may pave the way towards their formulation. So the credibility of global models describing Earth dynamics in a younger epoch like the late Quaternary is certainly not impaired by being in harmony with the ideas of Lovelock and Margulis. The testability of Gaian hypotheses, in turn, has been questioned from the outset (Kirchner 1989), however. The emerging field of astrobiology, investigating the cosmic odds of life through inspection of extra-solar planets, now introduces certain elements of reproducibility into Gaia theory and thereby generates some hope regarding the ultimate construction of selective criteria for 'geophysiologic' principles.

Within the last years widely recognized breakthroughs in astronomical measurement techniques have provided overwhelming evidence for the existence of extra-solar planets. These techniques exploit both the gravitational influence exerted by every planet on its more massive central star and the fractional reduction in the starlight during planetary transition (see e.g. Marcy *et al.* 2000). Up to now more than eighty planets outside the solar system have been detected (Schneider 2002). Most of these celestial bodies seem to have a rather huge mass and unexpectedly small orbits. But there have also been reports of the discovery of an Earth-sized planet in the centre of the Milky Way (Rhie *et al.* 1998).

This astrophysical progress has stimulated new speculations and assessments concerning the habitability of cosmic entities. One of the dominant concepts involved here is the so-called habitable zone (*HZ*), which is defined as the range of orbital radii R with respect to a given central star where an Earth-like planet might enjoy sufficiently moderate surface temperatures as required for the development of advanced life forms. The HZ for any star–planet pair can be estimated using integrated global models like the one constructed and operated at the Potsdam Institute (Franck *et al.* 1999, 2000*a*, *b*). This model couples formal representations of the solid Earth, the atmosphere, the hydrosphere, and the biosphere, and simulates planetary evolution under external forcing like increasing central-star luminosity.

Our model-based search for extra-solar habitability has already generated a number of results, although no single planetary candidate has passed the test yet. We employ a photosynthesis-related definition of the habitable zone, namely,

$$HZ(t) = [R_{\text{inner}}(t), R_{\text{outer}}(t)] = \{R \mid) (R(t)) > 0\}, \tag{1}$$

where) is an appropriate bioproductivity function and t denotes geological time. In Franck *et al.* (2000*b*) this HZ is calculated dynamically for extra-solar planetary systems using the luminosity evolution of central stars on the so-called main sequence in the mass range between 0.2 and 2.5 solar masses (M_s). As an illustration, we present the results for $M_s = 0.8$, 1.0, and 1.2, respectively, in Figure 1.1.

If we assign the name Gaia to the Earth's ecosphere in the sense of a global self-organized organismic machinery, then we can ask the question about the existence and abundance of 'sister Gaias' in our galaxy. Let N_{Gaia} denote the number of such contemporary relatives of our planet in the Milky Way. Using a number of assessment tools including our simulation model, N_{Gaia} can be estimated in a no-nonsense way. In fact, N_{Gaia} may be derived from a subset of factors concatenated by the famous Drake Equation (see e.g. Dick 1998; Jakosky 1998; Terzan and Bilson 1997):

$$N_{\text{Gaia}} = N_{\text{MW}} \cdot f_{\text{p}} \cdot n_{\text{CHZ}} \cdot f_{\text{n}}. \tag{2}$$

Let us discuss the individual factors in some detail now: N_{MW} is simply the total number of stars in the Milky Way. This figure is quite well known to be close to $4 \cdot 10^{11}$ (see e.g. Dick 1998). The next factor, f_{p}, is the fraction of stars with planets—an astrophysical entity entirely speculative only a decade or two ago. Recent findings indicate that planetary systems may be common for stars like the Sun, at least. Current extra-solar planet detection methods are sensitive to giant planets only. According to Marcy and Butler (2000) and to Marcy *et al.* (2000), approximately 5 per cent of the Sun-like stars surveyed so far possess giant companions. This means, of course, that our solar system is not at all typical. In order to estimate the galactic fraction of stars with *Earth-like* planets as relevant for ecosphere

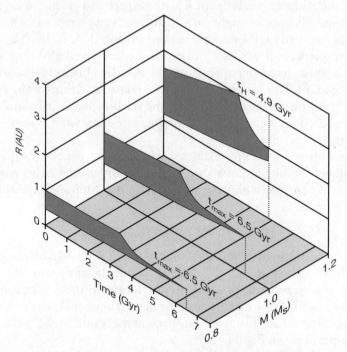

FIG. 1.1. Width and position of the *HZ* as determined by the Potsdam geo-dynamic model for three different stellar masses, i.e., $M = 0.8$, 1.0, and $1.2\ M_s$. The orbital radius is measured in astronomical units (AU), and geological time in gigayears. t_{max} denotes the maximum life span of the planetary biosphere as determined by processes like weathering and plate tectonics. τ_H denotes the hydrogen burning time on the main sequence limiting the astrophysical lifetime of more massive stars.

hunting, a number of theoretical considerations have to be employed: Lineweaver (2001) combines star and Earth-sibling formation rates based on the metallicity of the host star, for instance. In astrophysics metallicity is defined as the ratio of all elements heavier than helium ('metals') to hydrogen. These 'metals' seem to be necessary for our existence because they are constituents of both Earth-like planets and of all forms of life known up to now. This consideration can be directly used for approximating f_p. As a rather conservative estimate, we pick the value 0.01, which refers—as mentioned above—to Earth-like planets already and is certainly much smaller than the general planet–star ratio.

The third relevant factor, n_{CHZ}, is the average number of those planets per appropriate planetary system which stay continuously (i.e. several billion years) in the *HZ*. Our simulation modelling approach allows us to calculate this number in a rather reliable way (see Fig. 1.1). As demonstrated by Franck *et al.* (2001), $n_{CHZ} = 0.012$ for the most advanced geodynamic model.

Finally, f_n is the probability that life really develops on a planet that is in the habitable zone. In other words, these are the planets supporting Gaia as a sophisticated form of cosmic evolution. The magnitude of f_n has been a topic of controversial discussions for many years now. The crucial question is whether the first principles of chemistry necessarily lead to replicating molecules on habitable celestial bodies. Some scientists hold that in the presence of liquid water, carbon, and certain basic nutrients simple microbial life is almost certain to develop (Marcy 2001). This is equivalent to suggesting that f_n is close to 1: if life can happen, it does (see e.g. Dick 1998). Others, however, argue that f_n must be an extremely small number (Hart 1995). We pick the more conservative value $f_n = 10^{-2}$ to counteract the more optimistic choices made for the other factors. Thus we finally arrive at

$$N_{Gaia} \approx 4.8 \cdot 10^5, \tag{3}$$

which is, in fact, a rather big number and indicates that Gaias should be abundant in our galaxy. It is evident, though, that each of the four factors in Equation 1 is associated with large error bars. In this sense, our result must be perceived as a thoroughly educated guess.

The famous question 'Is Earth commonplace in the Milky Way?' can be answered 'Yes!' according to the considerations made above. This may be still premature, however, as our estimation does not include important preconditions or perturbing factors (like the presence of a large moon, the company of a giant planet, the abundance of long-lived radioisotopes, or the occurrence of destructive cosmic events) that can significantly reduce the number of sister Gaias (Franck *et al.* 2001).

The ultimate question goes one step further and asks 'Are we the only intelligent species in our galaxy?' This riddle is addressed by the SETI initiative (Search for ExtraTerrestrial Intelligence with the help of radio-signals) and formalized by the full Drake Equation (see e.g. Drake and Sobel 1992), i.e. Equation 2 with two additional factors: the first one is the fraction of Gaia's sisters developing technical civilizations; the second is the average ratio of civilization lifetime to Gaia lifetime. The latter entity actually dominates everything else; its value may be as small as 10^{-7} (Franck *et al.* 2001). So the number of civilizations whose radio emissions might be detectable could be really minute, if not equal to zero.

TOWARDS A SYSTEMS ANALYSIS OF THE EARTH

In the previous considerations we have used the notion 'Earth System' as a convenient envelope without making any formal attempt to specify its content. Let us try now to develop a definition which is less vague than the standard connotation of this increasingly popular term.

At the highest level of abstraction, the make-up of the Earth System $\mathcal{E}$ can be represented by the following 'equation':

$$\mathcal{E} = (\mathcal{N}, \mathcal{H})$$
$$\downarrow \quad \downarrow$$
$$(a,b,c,..) = \mathcal{N}\,\mathcal{H} = (\mathcal{A}, \mathcal{S}) \tag{4}$$
$$\downarrow$$
$$\mathcal{S} = (\mathcal{B}, \mathcal{V}, \mathcal{O}).$$

A detailed interpretation of this formal structure is given in Schellnhuber (1998); here we restrict ourselves to sketching the crucial features only. First of all, Equation 4 expresses the elementary insight that the overall system in question consists of two main components, namely the ecosphere $\mathcal{N}$ and the human factor $\mathcal{H}$. $\mathcal{N}$ consists, in turn, of an alphabet of intricately linked planetary sub-spheres a (atmosphere), b (biosphere), c (cryosphere), and so on. This is the entity often described using the metaphor 'Gaia' (as in the previous section). The human factor is even more subtle: $\mathcal{H}$ embraces the 'physical' sub-component $\mathcal{A}$ (*anthroposphere* as the sum of all individual human lives, actions, and products) and the 'metaphysical' sub-component $\mathcal{S}$ reflecting the emergence of a Global Subject. This subject is a self-organized co-operative phenomenon, a self-conscious force driving global change either to sustainable trajectories or to self-extinction. For the time being, let us just mention that $\mathcal{S}$ may be decomposed into three constituents, namely $\mathcal{B}$ (the global 'brain'), $\mathcal{V}$ (the global value system, or 'soul'), and $\mathcal{O}$ (the executive organs of the Global Subject). We will return to this 'metaphysics' later in this essay.

The emergence of Earth System science is nicely reflected by the fact that major international research programmes either exist already or are being established now for the investigation of all main factors in Equation 4, i.e. $\mathcal{N}$, $\mathcal{A}$, and $\mathcal{S}$. No less than three such programmes are dedicated to the ecosphere, namely the World Climate Research Programme (WCRP, see WCRP 2001), the International Geosphere–Biosphere Programme (IGBP, see Steffen and Tyson 2001), and DIVERSITAS, a scientific network *in statu nascendi* that will unify various approaches to the study of biodiversity (see ICSU 2001). WCRP is the best established and best funded enterprise within this troika, and has already advanced the quantitative understanding of the physical climate system beyond the expectations expressed when it was launched more than twenty years ago.

The most integrative and systemic programme, though, is the IGBP, which was created in the 1980s with the objective of describing and understanding the interactive physical, chemical, and biological processes that regulate the total Earth System, the unique environment that it provides for life, and the changes that are occurring in this system. Table 1.1 yields an impression of the complexity of the IGBP mission by summarizing its so-called 'core projects'. Most of these activities are devoted to truly global topics. The BAHC project, for instance, tries to determine in quantitative terms how the planetary hydrological cycle is influenced by the terrestrial biosphere through interception, evapotranspiration, retention, and other fundamental processes (Kabat *et al.* forthcoming).

IGBP is now in its synthesis phase, epitomized by the Amsterdam summit mentioned at the beginning, and will be relaunched in 2003 as a full Earth System science programme with appropriate targets and structures. When it comes to devising and managing this transition, a crucial role will be played by the GAIM Task Force mentioned in Table 1.1, which is expected to develop, much like a think-tank, the overall vision for the new IGBP. GAIM's present and future activities are guided by the Waikiki principles (Schellnhuber 2000), summarizing the main tasks as follows:

1. Exploration of trans-project research needs and opportunities.
2. Conceptual and methodological integration of wisdom inside and outside IGBP.
3. Advancement of Earth System modelling at all adequate levels of complexity.

TABLE 1.1. *Present main activities of IGBP*

Acronym	Issue
BAHC	Biospheric Aspects of the Hydrological Cycle
DIS	Data and Information Service
GAIM	Global Analysis, Integration, and Modelling
GCTE	Global Change and Terrestrial Ecosystems
GLOBEC	Global Ocean Ecosystem Dynamics
IGAC	International Global Atmospheric Chemistry
JGOFS	Joint Global Ocean Flux Study
LOICZ	Land–Ocean Interactions in the Coastal Zone
LUCC	Land Use/Cover Change
PAGES	Past Global Change
SOLAS	Surface Ocean–Lower Atmosphere Study
START	Global Change System for Analysis, Research, and Training

Most important is, in fact, the third principle: making Earth System science really happen demands an extensive toolkit comprised of powerful techniques that either refine and extend existing approaches or open up completely new ways of investigation. A vast array of mathematical models is now available for simulating the dynamics of the total ecosphere or of parts of it. Stylized models focus on the description and understanding of the major features of the planetary machinery only, while comprehensive 'Earth Simulators' are being assembled from the most sophisticated component modules. In between, Earth System models of intermediate complexity (EMICs) have already proven to be effective tools for both hindcasting and forecasting ecosphere behaviour.

At present the EMICs road to systems-level analysis of the global environment appears to be the most promising one. These models describe most of the processes implicit in the full-complexity models, albeit in a more parameterized form. They nevertheless simulate the interactions among several, or even all, components of the ecosphere explicitly. On the other hand, EMICs are simple enough to allow for long-term simulations over several periods of 10,000 years or a broad range of sensitivity experiments with the digital caricatures of the planet. Fig. 1.2 exhibits the structure of CLIMBER-2, an EMIC developed and extensively employed at the Potsdam Institute. Meanwhile, the worldwide pool of EMICs has grown to more than ten items and contains such diverse simulation machines as the Bern 2.5D Climate Model and the MIT Integrated Global System Model (Claussen 2001). The further expansion and maturation of the EMIC family is one of GAIM's top priorities in the medium term.

Incidentally, Figure 1.2 provides a telling illustration of the quite unbalanced scientific representation of natural and human dimensions, respectively, in present planetary dynamics simulators: the entire anthroposphere including its industrial metabolism appears as a harmless and faceless manikin. This is an adequate reflection of the fact that our quantitative mimicry skills for global civilizatory processes like transmigration dynamics, accelerated urbanization, or international technology diffusion lag far behind our ecosphere-simulation capacity. There are vital signs of hope yet, as the International Human Dimensions Programme on Global Environmental Change (IHDP, see IHDP 2001) is now initiating a number of ambitious research projects dedicated to the '$\mathscr{A}$ factor'. IHDP is, in particular, focusing on the causes and consequences of people's individual and collective actions affecting the planet's life support systems.

Owing to these and similar efforts, reasonable—if not necessarily valid—models of large-scale socio-economic dynamics may be available for coupling to appropriately designed ecosphere simulators in a few years from now. Such a direct coupling has been attempted so far only in the so-called

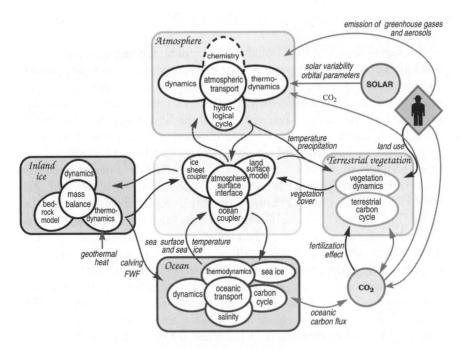

FIG. 1.2. The structure of the CLIMBER model, an example for an Earth System model of intermediate complexity. A full description is provided by Petoukhov *et al.* (2000) and Ganopolski *et al.* (2001).

Integrated Assessment Models for climate change analysis (Schneider 1997; Smith *et al.* 2001), but their trustworthiness has been rather limited due to the persistent dominance of natural-science ingredients.

The ultimate challenge, however, is the '$\mathscr{S}$ factor', i.e. humanity's collective consciousness and activity as a planetary force that can *decide*, to a certain extent, on the future global environment. The basis for the creation of a Global Subject is worldwide transportation and communication that generate, i.e. thousands of virtual societal super-cells such as personal networks, pressure groups, non-governmental organizations, epistemic communities, and global corporations. The decisive ingredient in the making of $\mathscr{S}$, however, is added by science that allows this global actor to watch itself in action (e.g. opening a stratospheric ozone hole over Antarctica) and thus establishes the peculiar quality of self-referentiality. Or, as Jolly puts it: '*Homo sapiens* is slowly evolving into something akin to a superorganism, a highly-structured global society in which the lives of everyone on the planet will become so interdependent that they may grow and develop with a common purpose. It may take centuries to fully evolve this global creature, but there is no question that it will . . .' (Jolly 1999; see also Rosnay 1995).

The supreme task of the Global Subject will be the selection and implementation of a tolerable environmental future from the infinity of optional co-evolutions of ecosphere and anthroposphere. In other words, $\mathscr{S}$ must guarantee *sustainable development*. This might be achieved, for instance, by international agreements like the Kyoto Protocol under the Framework Convention on Climate Change, but it is not clear at all whether isolated measures of that kind will really add up to a comprehensive global environmental management strategy. In fact, a new type of science is required to explore the features, objectives, and instruments of $\mathscr{S}$ as resulting from the volition of billions of individuals. This enterprise may be called 'Sustainability Science' (Kates *et al.* 2001) and will focus on the (possibly) irreversible responses of the nature–society system to multiple and interacting stresses. Combining different ways of knowing and learning will ultimately permit social actors to work in concert at all spatial levels of organizations, even with much uncertainty and limited information. Sustainability Science is not yet a fully fledged research programme like WCRP, but its further development is indispensable for the intellectual closure of Earth System analysis in the sense of Equation 4.

GAIA'S PAST VARIABILITY

The contemporary role of humanity in the Earth System can be assessed in view of the Earth's historical functioning, and one of the major tasks for Earth System science is to operationalize such an assessment. Fortunately, in addition to new scientific frameworks which need to be developed, Earth System science can draw on an ever richer pool of disciplinary tools, not only from the traditional Earth sciences, but also from systems theory and non-linear dynamics. In the following paragraphs some of these tools will be illustrated through examples in relation to characteristic features of Gaia.

Response of the Non-Linearly Coupled Climate System to Astronomical Forcing

In the early days of non-linear dynamics, phase-space reconstruction (Nicolis and Nicolis 1986) appeared as a promising tool to extract system properties from time series. However, for 'real', i.e. noisy, geophysical data, this may prove difficult (Grassberger 1986). An evaluation using bispectral statistics of a given time series is a more robust method to detect non-linear properties of the underlying system. With this tool Hasselmann (1963) detected non-linear wave mixing in a time series taken in shallow water. In a recent application of this technique Rial and Anaclerio (2000) analyse the Vostok ice core record (Petit *et al.* 1999) and add support to Milankovitch's (1969) theory. His theory would allow an interpretation of such a time series as the climate's response to astronomical forcing. At first glance, it could be seen as a weak point of Milankovitch's theory that the Fourier spectrum of the Vostok ice core's time series reveals peaks at 29,000 years ago (ka) and 69 ka, both of which are *not* present in the astronomical forcing. However, bispectral analysis showed (Rial and Anaclerio 2000) that both frequencies can be interpreted as phase-coherent sidebands from the (astronomical) 41 ka obliquity signal which are generated by frequency modulation with an overtone of the 413 ka eccentricity. This adds support to the hypothesis that the climate signal is indeed a non-linear response to astronomical forcing. Rial and Anaclerio (2000) stress that the importance of the Vostok ice core record lies not only in its unprecedented length but also in the way it was exploited. It is—in contrast to most deep-sea sediment records—*not* tuned to the 41 ka obliquity. It is not possible to observe a frequency modulation of 41 ka in a time series tuned to the obliquity, as the tuning—by definition—removes the effect of the modulation. This demonstrates that non-astronomic dating methods for palaeo-records are highly desirable and stresses the need for concerted inter-action—if not mutual education—of the various Earth System science branches.

Dansgaard–Oeschger Events

Presumably, it is not a mere coincidence that agriculture and hence modern civilization developed over the last 10,000 years, during a phase of climatic stability known as the Holocene. By contrast, in the 100,000 years that preceded the Holocene, the climate was highly variable and switched abruptly between cold and warm modes with a dominant periodicity of some 1,500 years (as displayed, for example, in Greenland ice core data; Grootes *et al.* 1993). The temperature in Greenland changed by up to 10°C in a matter of decades; these changes are known as Dansgaard–Oeschger events. In Ganopolski and Rahmstorf (2001) a mechanism is proposed which links the Dansgaard-Oeschger events to switching between competing operational modes of the thermohaline circulation (Rahmstorf 2000; Stommel 1961) in the North Atlantic. The authors utilize two key concepts of dynamical systems theory: the coexistence of multiple stable equilibria (equilibria are steady states of the underlying dynamics) and stochastic resonance.

The metaphor underlying the thermohaline circulation is that of a conveyor belt, which was introduced by Broecker and Peng (1982) as part of a model describing deep-sea circulation in large-scale oceanography. In Brüning and Lohmann (2000), for this example, the impact of scientific metaphors on the creative process within science is emphasized, and one may speculate as to which other Gaia-type metaphors may be 'activated'. While there was some discussion in which sense the conveyor belt actually exists, today there is no doubt that the thermohaline circulation makes a major contribution to the heat budget of the North Atlantic region (Rahmstorf 2000). The conveyor belt metaphor stimulated, furthermore, a series of ocean box models which emphasize different aspects of the thermohaline circulation (Titz *et al.* forthcoming). Systematic bifurcation analyses have been performed yielding a spectrum of scenarios for a breakdown of the thermohaline circulation. One can infer as certain that the amount of fresh water entering the North Atlantic is a key control parameter. Bifurcation diagrams display parameter combinations which allow for an equilibrium solution of the assumed dynamical equations. At a bifurcation point a qualitative change occurs, usually by the birth or the disappearance of a solution. The original Stommel model (1961) contains a simple positive feedback—overturning enhances salinity owing to salt transport from the south, and high salinity in turn enhances overturning—which allows for either two ('on' and 'off' mode) stable solutions, or one stable solution. The two regimes are connected by a saddle–node bifurcation. Beyond this 'critical threshold' of the control parameter, only the off-mode exists. For some box models, in addition, a *Hopf* bifurcation occurs (Titz *et al.* forthcoming). This bifurcation has consequences for the basin of attraction, and hence the robustness of the on mode. Such additional features were not found in more comprehensive two-dimensional ocean models—a discrepancy which still needs to be understood. However, reduced form models primarily should be regarded as a source of inspiration and guidance but not be utilized for prognostic purposes—at least as long as they are not rigorously *deduced* from the full equations of motion.

The present-day climate state is characterized by a warm (switched-on) mode. Although a second stable state exists for the present-day climate representing a cold mode (switched-off thermohaline circulation), a transition between the two has not occurred during the Holocene because of the relatively large basin of attraction of the warm mode in parameter space. On the contrary, during the last glacial, a stable (cold) and a metastable (warm) mode existed for the dynamics of the thermohaline circulation with a much smaller basin of attraction for the cold mode. Utilizing the climate model of intermediate complexity CLIMBER-2.3, Ganopolski and Rahmstorf (2001) showed that a relatively small perturbation of the freshwater input at high latitudes would be sufficient in order to switch the system into the metastable state whose lifetime was found to be of the order of several hundred years. A sinusoidal modulation with amplitude much smaller than typical system threshold values would not induce switching in present-day climate but it would result in periodic switching under glacial conditions. Ganopolski and Rahmstorf (2002) indicate that in the presence of noise, this driving amplitude can be further reduced, resulting in a flipping behaviour, synchronized with the harmonic forcing, yet missing one or the other beat. Such behaviour is known as stochastic resonance (Gammaitoni *et al.* 1998).

In summary, Dansgaard–Oeschger events can be understood as comparably fast switching events between coexisting stable equilibria, induced by a weak periodic forcing of the thermohaline circulation in addition to noise. This external trigger becomes amplified owing to the coexistence of multiple (meta)stable states of thermohaline dynamics that are mathematically impossible in a purely linear system. In the next section additional examples will be outlined which further establish the coexistence of non-linearity and multiple equilibria as a central concept in Earth System science. In investigating the fragility of the Earth System, one feels tempted to identify fragility within a dynamical system's framework as the chance to force the Earth System out of the Holocene's quasi-equilibrium either by anthropogenic influence or by internal fluctuations.

GAIA'S HOLOCENE MODE OF OPERATION

The Parts of Gaia

Before we further exemplify techniques which deal with pertinent Earth System processes, we want to sketch how Gaia's theory was further developed. As the common scientific toolbox typically deals with the parts of a system and their mutual relations, Volk (1997) identified such parts of Gaia. He suggested the following seven decompositional taxonomies (which he called 'viewpoints'), each with advantages and shortcomings: biomes, trophic guilds (producer, detrivore, carnivore, herbivore), biogeochemical cycles, primary substances (air, ocean, soil, life), genetic domains, and eukaryotic kingdoms.

Margulis and Schwartz (1988) use cell type as a fundamental criterion for classification which leads to the definition of 'domains'—eukaryotes and prokaryotes (cells with and without nuclei, respectively). While the taxonomy of domains might be justified, as domains mark types of genetic innovation, this taxonomy turns out to conflict with more functional properties. This phenomenon of mutual incompatibility holds for any pair of the above taxonomies, an observation that led Volk (1997) to keep the full collection of viewpoints as the best description of the whole.

Keeping the full collection of viewpoints defines a meta-method which is to be preferred as long as various descriptions are not reduced to a fundamental theory. The same discussion can be followed in the climate modelling community where (comprehensive) general circulation models (GCMs) have adopted a superior position. However, as suggested in Shackley *et al.* (1998) and more recently by the Non-Linear Initiative of IGBP (Pielke *et al.* forthcoming), the full hierarchy of models (from comprehensive to conceptual models) is necessary for an adequate understanding of the Earth System.

In the following paragraphs we present a view on Gaia which emphasizes an 'inspectory' view on the Earth System. Rather than describing the outputs of models with highest spatio-temporal resolution in detail, we want to highlight prominent qualitative features which we expect to be useful for a conceptual understanding of the ecosphere as not only an interconnected system, but also a structured entity.

Teleconnections

In the first approximation of ecosphere dynamics, consequences of actions exerted somewhere to the system result in local, or (on average) diffusive global effects. However, since the early 1930s the existence of large-scale recurrent atmospheric spatial patterns has been recognized, stressing a global, yet spatially structured interconnectedness of the ecosphere. Walker and Bliss (1932) identified the North Atlantic and the North Pacific Oscillations. Later the Arctic Oscillation, the North Pacific–Pacific/North American Pattern, and the Southern Oscillation were also recognized. For these large—if not global—patterns, Wallace and Gutzler (1981) redefined the term 'teleconnections'. We would like to include other types of large-scale phenomena which refer to the oceans or combined atmosphere–ocean patterns, such as the thermohaline circulation of the Atlantic and the El Niño/Southern Oscillation (ENSO) (Neelin *et al.* 1998). The concept of teleconnections is primarily descriptive; reminiscent of the 'parts of Gaia'. We will use them to exemplify further conceptual, systems-analytic tools.

The thermohaline circulation is a relatively isolated phenomenon which can be successfully addressed by tools from non-linear dynamics theory, and has already been described in some detail in the previous section.

ENSO represents an interplay of coupled ocean–atmosphere phenomena with a predominant period of approximately four years. El Niño is the warm phase of ENSO, whereby a weakening of the prevailing easterly trade winds in the equatorial Pacific allows the eastward propagation of warm waters. The latter normally accumulate to the west of the Pacific basin—with profound economic consequences, especially for South America. ENSO modelling provides an example where competing methodologies (Johnson 1999) yield similar results. One hypothesis states that an unstable mechanism exists in the coupled dynamics of the tropical Pacific ocean–atmosphere system (Bjerknes 1969). In that spirit, many models which contain at least one growing eigenmode (usually identified as ENSO) have been utilized to make skilful predictions of sea-surface temperature anomaly with lead times up to a year (Chen *et al.* 1995). A competing school of modellers follow Penland and Magorian (1993) and assume only stable (decaying) eigenmodes of ENSO, driven by stochastic forcing; hence, without weather there would be no El Niño events (Johnson 1999). We regard it as quite remarkable that empirical stochastic models are still of roughly the same predictive skill (Johnson *et al.* 2000) as state of the art GCMs. One can conclude that either ENSO is a linear stochastic phenomenon, or GCMs are currently missing important processes or optimized parameterizations.

Finally we would like to mention a fascinating speculation (Barnett *et al.* 2000) of a global hypermode which links three of the above teleconnections—the North Atlantic Oscillation, the Southern Oscillation, and the North Pacific Oscillation—to the monsoon. There seems to be empirical evidence of phase synchronization (Rosenblum *et al.* 1996) which is switched on and off on decadal time scale; i.e. the hypermode is not permanent. For this reason, phase synchronization has evaded detection by standard correlation techniques so far. If this hypothesis can be validated, it would be yet another success of an approach which tries to dynamically explain semi-separable 'modes' of Gaia with tools from non-linear dynamical systems theory.

Switch and Choke Points

While the concept of teleconnections emphasizes the global interconnectivity of the Earth System, at least within certain modes, we would like to relate also global to local dynamics. We ask whether there exist rather confined regions—choke points—with a strong influence on large-scale dynamics. If, in particular, this influence is related to rather distinct modes of operation between which fast transitions are possible, we prefer for such a choke the term 'switch point'. For the anthroposphere, the world's capitals, the stock market locations, and further hot spots of innovation (e.g. Silicon Valley) are undoubtedly such locations. (Note that in general, one would expect anthropospheric choke points to be located in developed countries which notably are susceptible to terrorist attacks.) In the spirit of the previous chapter, we will focus now on—subjectively selected—examples from the ecosphere (see Fig. 1.3).

While today there does not exist any ENSO model with predictive skill beyond one year, that does not rule out the possibility to control ENSO, at least within the physics of the (ENSO) Zebiak–Cane model (Zebiak and Cane 1987). Tziperman and Scher (1997) identified a choke point at the western boundary of the Pacific Ocean. According to their conclusions for this region, well-defined perturbations cannot suppress ENSO but they make it periodic and, hence, predictable. The underlying idea includes a delay-coordinate phase space reconstruction, an identification of the dominant unstable periodic orbit related to ENSO, and controlled 'kicks' of the dynamics from the unstable into the stable manifold of that orbit. However, the question arises whether this mechanism would still persist in a more realistic model which includes noise.

As an example of atmospheric physics, the onset of Indian monsoon seems to be susceptible to snow coverage of the Tibetan plateau (Webster *et al.* 1998). Further, the monsoon impacts on vital atmospheric elements like Asian agriculture are extremely susceptible to the onset and extent of monsoon. Additionally, the Indian monsoon seems to affect ENSO and vice versa. Lag correlation relationships suggest that a strong (weak)

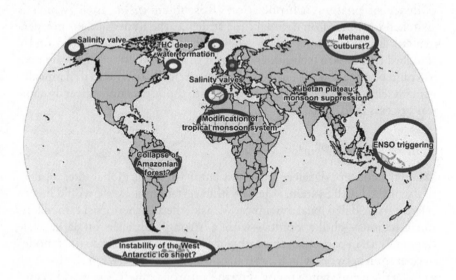

FIG. 1.3. Possible switch and choke points of the Earth System. Changes at the indicated regions may have a large influence on the whole system or major parts of it.

summer monsoon tends to lead to a La Niña (El Niño) in the later seasons of the year. And, descriptively, all El Niño years in the Pacific Ocean were followed by drought years in the Indian region between 1871 and 1991.

One of the most prominent examples of a switch point refers to those regions where the thermohaline circulation is most susceptible to freshwater forcing, leading to a complete shutdown in the extreme case. According to Rahmstorf (1996), model simulations suggest that the regions of North Atlantic deep-water formation, i.e. the Labrador and the Greenland Sea, can be identified as switch points.

It is also well established in ecology that non-linearities in biochemical and biophysical processes can result in the existence of several steady states, especially on a local scale (Scheffer *et al.* 2001): for example, in some regions one stable ecosystem (forest) can be replaced by another stable ecosystem (grassland) after a severe fire event. Alteration of vegetation cover leads to changes in the local climate, creating a feedback loop between vegetation and climate. In subtropical deserts, such as the Sahel–Sahara region, the atmospheric–vegetation feedbacks can be strong enough to influence the atmospheric dynamics even on the continental scale. Analysis of aeolian sediments off the coast of West Africa shows a rapid termination of the African humid period about 5,500 years ago (Menocal *et al.* 2000). This dramatic change, which is presumably related to an abrupt dieback of Saharan and Sahelian vegetation, occurred within several decades, i.e. extremely fast in comparison with the slow changes associated with orbital forcing. Theoretical considerations and numerical experiments (Claussen *et al.* 1999) suggest that the abrupt desertification in North Africa could be explained as an interplay of internal, mainly regional, vegetation atmosphere feedbacks and an external trigger by orbital forcing. As regards thermohaline circulation, these feedbacks can lead to multiple equilibrium states with the possibility of abrupt changes when thresholds are crossed (Brovkin *et al.* 1998). Similar considerations are applicable to Sahelian rainfall variability (Wang and Eltahir 2000).

Another type of switch point might be given by clathrate destabilization. Clathrates (gas hydrates) are naturally occurring ice-like solids in which water molecules trap gas molecules in a cage-like structure. They can be found in the oceanic sediment of continental slopes, in the deepwater sediment of inland lakes and seas, and in permafrost areas of polar regions. The amount of carbon stored in methane hydrates worldwide is estimated to be in the order of twice the amount contained in all fossil fuels on Earth (Kvenvolden 1999; Gornitz and Fung 1994). Since, integrated over a hundred-year time span, the warming potential of methane is twenty times larger than an equivalent weight of carbon dioxide, methane hydrate deposits can be viewed as climate switch points due to the prospect of their facilitating a runaway greenhouse effect. This is by no means pure speculation as the huge carbon isotope ratio $^{13}C/^{12}C$ anomaly in palaeo-records during the Palaeocene–Eocene Transition, about 55 million years ago, are explained best by scenarios of methane hydrate destabilization (Dickens 1999; Norris and Röhl 1998). This release is correlated to a dramatic warming of high-latitude surface water and deep water by 5–7°C. Importantly, the magnitude of carbon emission rates at that time is estimated to be comparable to the rates of current anthropogenic carbon input. Concerning present-day climate, Mienert *et al.* (1998) assume for an investigated shallow-water gas hydrate site in the Barents Sea that a temperature rise of just 1°C might be sufficient to move the reservoir out of its stability regime.

On a more local scale another impact can be generated by additional switch properties of gas hydrates. Huge amounts of methane can be released by submarine landslides when the gas hydrates of sediments dissociate. A prominent example of an enormous slide some thousand years ago is the Storrega avalanche off the coast of Norway (Mienert *et al.* 1998). On the Scottish coast a tsunami brought about by one of these slides (6–8 ka), deposited sediment up to 4 metres above high-water mark.

A global and permanent sea-level rise of several metres, on the other hand, might occur through a completely different mechanism, e.g. by the destabilization of the West Antarctic Ice Sheet (Ahmad *et al.* 2001).

Finally, we would like to mention the existence of salinity valves, with regard to which some debate has occurred on whether the construction of bridges or other obstacles would severely alter the salinity exchange rates operating, for example, in the Bering Strait (Aagaard and Carmack 1989), the Strait of Gibraltar (Rahmstorf 1998), and the Skagerrak (Omstedt and Axell 1998). Rahmstorf (1998), for example, rejects the bizarre call by Johnson (1997) for a dam across the Strait of Gibraltar to prevent a new Ice Age, indirectly caused by the Aswan Dam. Even within a mono-causal paradigm, the suggested measure would be inadequate as the anticipated mechanism turned out to be exaggerated. On the other hand, there is little doubt about the importance of the Skagerrak for various biogeochemical exchange rates at the mouth of the Baltic Sea.

ANTHROPOCENE FORCING OF THE EARTH SYSTEM

Humankind is currently inducing changes to the Earth System which are unprecedented in human history, with respect to both magnitude and rate. Accordingly, Crutzen and Stoermer (2000) coined the term 'anthropocene' for the present epoch, which they consider to begin with the onset of the industrial revolution. They support this notion—which has quickly been adopted in the scientific community—by a list of Global Change phenomena, of which the most prominent might be the creation of the Antarctic ozone hole. In this case, mankind was lucky in that it was able to correct the trend through the Montreal ozone protection protocol. If, however, 'the chemical industry had developed organobromine compounds instead of CFCs then . . . we would have been faced with a catastrophic ozone hole everywhere and all year round' (Crutzen 1996). The other phenomena Crutzen and Stoermer (2000) list fall into three categories, of which we mention briefly the most relevant examples. The first category refers to 'transformations of the biosphere' and concerns the loss of 50 per cent of mangroves in coastal wetlands, the conversion of 30–50 per cent of the total land surface, and finally the sixth great extinction event in Earth's history. The anthropocene indeed qualifies for an extinction event, as the species extinction rate is amplified by a factor of 10^3–10^4. A second category, 'transformation of biogeochemical cycles', concerns the enhancement of greenhouse gas concentrations (CO_2 by 30 per cent, CH_4 by 100 per cent), and the facts that human emissions have become comparable to natural ones (SO_2 more than 200 per cent, NO more than 100 per cent of the natural rate), and that humans fix more nitrogen (mostly used for fertilizers) than all terrestrial ecosystems together. Finally, under 'exhaustion of resources', it is stated that currently more than 50 per cent of all accessible fresh water is being used, 25–35 per cent of fish primary production is being removed, and the fossil fuels aggregated over 100 to 1,000 million years will have been consumed within a period in the order of 100 years.

Which of the impacts of these rather dramatic changes are observed on an aggregated, physical level? According to the latest Intergovernmental Panel on Climate Change (IPCC) report (Ahmad *et al.* 2001), the global average surface temperature has increased by $0.6 \pm 0.02\,°C$ (95 per cent confidence) over the last century. Additional palaeo-information reveals that the rate and duration of warming has been much greater than in any of the previous nine centuries. The report claims that there is 'new and stronger evidence that most of the warming observed over the last 50 years is attributable to human activities'.

Crisper evidence of the unprecedented nature of mankind's influence on the global cycles, however, is delivered by the famous Vostok ice core data set (Petit *et al.* 1999). It reveals that anthropogenic emissions have pushed greenhouse gas concentrations far beyond the boundaries of the past 420,000 years, and has demonstrated how various environmental parameters are highly interdependent. It is now up to the IGBP community to put the consequences for the anthropocene into perspective. Is the climate system operating in a quasi-equilibrium, or will the current changes drive the ecosphere into new modes of operation in a runaway manner? If so, can the anthroposphere and the ecosphere adapt, or is it necessary and feasible to re-establish and stabilize pre-anthropocene conditions?

FACETS OF FRAGILITY

After highlighting the human factor as a major influence on the ecosphere, one feels tempted to ask whether there are immediate reasons for concern. European cultures have up to the present transformed almost 100 per cent of European landscapes, i.e. exerted a major forcing of the ecosphere as well, for their very well-being. What is the appropriate metric for an assessment?

The IPCC community has agreed on the global mean temperature as the common metric against which impacts are measured. The Earth is project-ed to warm by 1.4 to 5.8°C by 2100 relative to 1990 (Cubasch and Meehl 2001). While there is observational evidence that regional increases in tem-perature have already affected ecospheric systems, IPCC presents the 'Reasons for Concern' (Smith *et al.* 2001) on a qualitative scale with respect to

- the risks for unique and threatened systems;
- the risks from extreme climate events;
- inequitable distribution of impacts;
- dangerous aggregate impacts;
- risks from future large-scale discontinuities.

For a 5.8°C temperature increase, a high risk and negative impact for any of the above categories is projected, while for 1.4°C there is a (top-down) decreasing severity of impact to be expected, starting from 'risks to some systems' to 'very low risk from large-scale discontinuities'.

These considerations appear even more disturbing if new results from a fully coupled, three-dimensional carbon-climate model (Cox *et al.* 2000) are taken into account. Accordingly, under a 'business as usual' emission scenario, the carbon-cycle feedback turns the terrestrial biosphere into a carbon source after 2050. Within this scenario the Amazon rainforest plays a major role. While rates of photosynthesis increase with atmospheric CO_2 concentration, this effect is projected to become overcompensated by a dieback of most of the rainforest by 2100 (Cox 2001). The latter is sup-posed to be triggered by a reduction in precipitation associated with El Niño-like temperature patterns in the tropical Pacific. The reduction is enhanced by a positive feedback loop from forest via evapotranspiration to precipitation, making the Amazon forest a switch point (see above). The total carbon-cycle feedback mechanism is projected to lead to an additional global warming of 1.5°C by 2100, where the increase over land even amounts to 2.5°C.

In some areas of the world humankind will have to deal with a reduction in crop yield, decreased water availability, increased exposure to vector-borne diseases, and an increase in the risk of flooding, heavy precipitation, and sea-level rise. While especially the population in the mid-latitudes may, ironically, gain some benefit from global warming such as increased crop yield, one could ask whether the Earth System in its present state is fragile. In fact, the list of switch and choke points may provide some elements which could be activated by global warming. Precariously, a switch to new modes of operation is often accompanied by profound hysteresis (Scheffer *et al.* 2001), i.e. in order to restore previous modes of operation it is not enough to restore the related control parameters; on the contrary, a need for massive overcompensation is to be expected. Hence, the possibility that we are currently driving the *whole* Earth System into a new stable equilibrium with unknown properties should be a massive reason for concern.

WHAT SHALL WE DO?

In view of the last remarks it seems evident that some immediate action to reduce anthropogenic greenhouse gas emissions is necessary. Scientific necessities, however, do not generate appropriate political measures as a rule. The options for implementing sustainable developments depend in a most delicate manner on the value systems and decision-making procedures of modern societies. The dynamics of the '$\mathscr{S}$ factor' discussed in the third section is probably even more dominated by non-linearities than the operation of the ecosphere.

Yet a consensus paradigm for the co-evolution of humanity and nature during and beyond the twenty-first century must be chosen soon, as the manoeuvring space for global environmental management is closing down rapidly. In Schellnhuber (1998) the basic menu of sustainability principles has been assembled and formalized in some detail (see Table 1.2).

The *standardization* paradigm seeks to prescribe an explicit long-term co-evolution corridor; the *optimization* paradigm is concerned with maximizing an aggregated anthroposphere–ecosphere welfare function; the *pessimization* paradigm tries to avoid the worst even in the case of bad management by future generations; the *equitization* paradigm is focused on the preservation of options during co-evolution; and the *stabilization* paradigm intends to 'land' and maintain the Earth System in a desirable (or, at least, acceptable) state.

TABLE 1.2. *Taxonomy of pure sustainability paradigms*

Symbol	Paradigm	Positive goal	Negative motive
P_0	Standardization	Order	Despotism
P_1	Optimization	Prosperity	Greed
P_2	Pessimization	Security	Cowardice
P_3	Equitization	Fairness	Jaundice
P_4	Stabilization	Reliability	Indolence

Note: See Schellnhuber (1998) for a detailed derivation and description.

Elementary logic implies that the pessimization paradigm overrules all competitors: whatever goals humanity wishes to pursue in the course of co-evolution with the planetary life support system, our species must try to avoid pathways towards catastrophe domains in possibility space. Such domains are defined as both intolerable and inescapable (Schellnhuber 1998). The worst (not entirely fictitious) example for a catastrophe domain is a global climate regime where the runaway greenhouse dynamics would be triggered by self-amplifying geosphere–biosphere interactions. Less apocalyptic yet still unacceptable environmental changes that might be unleashed by civilizatory business-as-usual have been mentioned above—like the breakdown of North Atlantic deep-water formation, the transformation of the Asian monsoon system, or the intensification of the ENSO dynamics.

The pessimization paradigm can be operationalized into a sustainability strategy if 'guard rails' in co-evolution space are identified which separate the regular domains from the catastrophic ones. This approach has been demonstrated within the climate context through the 'tolerable windows concept' (Bruckner *et al.* 1999; Petschel-Held *et al.* 1999; Tóth *et al.* forthcoming). The basic idea is to confine the potential anthropogenic climate excursions to a subset with manageable impacts on natural and socio-economic systems. The German Advisory Council on Global Change (WBGU 1995), for instance, has proposed consideration of 2°C additional global warming and 0.2°C temperature increase per decade as critical values (guard rails) in climate phase space. The concept of preserving present equilibrium dynamics by appropriate management has been discussed recently in quite general terms (see e.g. Scheffer *et al.* 2001), certainly inspired by Holling's (1973) concept of ecosystems' resilience with respect to internal fluctuations and external perturbations.

In order to bring about sustainable development, however, the pessimization approach has to be completed by sub-strategies that ensure, in particular, equity and efficiency goals. In fact, the most reasonable strategy for planetary management (Newton 1999) seems to be a composite one, where the pessimization paradigm sets the boundary conditions, the equitization paradigm sets the agenda, and the optimization paradigm determines the best means to be employed. For the climate protection problem, this means observing the tolerable window in climate space by appropriate greenhouse gas emissions corridors, and adapting to the residual climatic change in the most cost-effective way as commensurable with first principles of justice (Rawls 1999).

Finally, as there is no single, once and for ever correct decision by whatever Global Subject (Schellnhuber 1998, 1999), a whole pool of robust adaptive strategies (reminiscent of 'fuzzy control') have to be considered. In other terms, 'policy prescriptions should avoid recommending strategic choices (including technologies) which are not compatible with a world in which policies and behaviours are forever being modified and adapted' (Morgan 2001). The Global Subject is actually a dynamic *process* supported by a well-informed public, perpetually debating the normative decisions involved via the World Wide Web and other communication platforms. While Earth System science cannot be normative itself, it must strive for an optimal interface between the cognitive basis required and the decision-makers, or stakeholders, in the widest sense. In fact, Earth System science will be self-adapting through its contribution to adaptive planetary management, so the intense interaction between existing scientific branches will not suffice (Steffen and Tyson 2001).

Quite to the contrary, this new type of science will be much more problem-orientated and contextual (Nowotny *et al.* 2001) than the traditional one, although the toolboxes assembled so far should not be scorned. While one might feel discouraged by the challenges Global Change imposes, one should not hesitate to realize the fascinating options involved for the further development of our common scientific enterprise.

REFERENCES

Aagaard, K., and Carmack, E. C. (1989). 'The role of sea ice and other fresh water in the arctic circulation'. *Journal of Geophysical Research*, 94 (C10): 14485–98.

Ahmad, Q. K., Anisimov, O., Arnell, N., *et al.* (2001). 'Summary for policy-makers', in J. J. McCarthy, O. F. Canziani, N. A. Leary, D. J. Dokken, and K. S. White (eds), *Climate Change 2001: Impacts, Adaptation and Vulnerability. IPCC Third Assessment Report.* Cambridge: Cambridge University Press.

Barnett, T., Timmermann, A., Pierce, D., and Schneider, N. (2000). 'The global hypermode'. Invited paper OS11D–08, American Geophysical Union Fall meeting, San Francisco.

Bjerknes, J. (1969). 'Atmospheric teleconnections from the equatorial Pacific'. *Monthly Weather Review*, 97: 163–72.

Broecker, W. S., and Peng, T.-S. (1982). *Tracers in the Sea.* Palisades, NY: Eldigio Press.

Brovkin, V., Claussen, M., Petoukhov, V., and Ganopolski, A. (1998). 'On the stability of the atmosphere–vegetation system in the Sahara/Sahel region'. *Journal of Geophysical Research*, 103 (D24): 31613–24.

Bruckner, T., Petschel-Held, G., Tóth, F. L., Füssel, H.-M., *et al.* (1999). 'Climate change decision-support and the tolerable windows approach'. *Environmental Modeling and Assessment*, 4: 217–34.

Brüning, R., and Lohmann, G. (2000). 'Charles S. Peirce on creative metaphor: a case study on the conveyor belt metaphor in oceanography'. *Foundations of Science*, 4: 389–403.

Chen, D., Zebiak, S. E., Busalacchi, A. J., and Cane, M. A. (1995). 'An improved procedure for El Niño forecasting: implications for predictability'. *Science*, 269: 1699–1702.

Claussen, M. (2001). 'Earth system models', in E. Ehlers and T. Krafft (eds), *Understanding the Earth System: Compartments, Processes and Interactions.* Heidelberg: Springer.

Claussen, M., Kubatzki, C., Brovkin, V., Ganopolski, A., *et al.* (1999). 'Simulation of an abrupt change in Saharan vegetation at the end of the mid-Holocene'. *Geophysical Research Letters*, 24 (14): 2037–40.

Cox, P. M. (2001). Private communication.

Cox, P. M., Betts, R. A., Jones, C. D., Spall, S. A., and Totterdell, I. J. (2000). 'Acceleration of global warming due to carbon-cycle feedbacks in a coupled climate model'. *Nature*, 408: 184–7.

Crutzen, P. J. (1996). 'My life with O–3, NOx, and other YZO(x) compounds (Nobel lecture)'. *Angewandte Chemie—International Edition*, 35 (16): 1758–77.

Crutzen, P. J., and Stoermer, E. F. (2000). 'The "Anthropocene"'. *IGBP Newsletter*, 4: 17–18 and 16.

Cubasch, U., and Meehl, G. A. (2001). 'Projections of future climate change', in J. T. Houghton, Y. Ding, D. J. Griggs, M. Noguer, *et al.* (eds), *Climate Change 2001: The Scientific Basis, IPCC Third Assessment Report.* Cambridge: Cambridge University Press.

Dick, S. J. (1998). *Life on Other Worlds.* Cambridge: Cambridge University Press.

Dickens, G. R. (1999). 'The blast in the past'. *Nature*, 401: 752–5.

Drake, F., and Sobel, D. (1992). *Is Anyone Out There? The Scientific Search for Extraterrestrial Intelligence.* New York: Delta.

Franck, S., Kossacki, K., and Bounama, C. (1999). 'Modelling the global carbon cycle for the past and future evolution of the Earth system'. *Chemical Geology*, 159: 305–17.

Franck, S., Block, A., von Bloh, W., Bounama, C., *et al.* (2000a). 'Reduction of biosphere life span as a consequence of geodynamics'. *Tellus*, 52B (1): 94–107.

Franck, S., Block, A., von Bloh, W., Bounama, C., *et al.* (2000b). 'Determination of habitable zones in extrasolar planetary systems: where are Gaia's sisters?' *JGR-Planets.* 105 (E1): 1651–8.

Franck, S., Block, A., von Bloh, W., Bounama, C., *et al.* (2001). 'Planetary habitability: is Earth commonplace in the Milky Way?' *Naturwissenschaften*, 88: 416–26.

Gammaitoni, L., Hänggi, P., Jung, P., and Marchesoni, F. (1998). 'Stochastic resonance'. *Review Modern Physics* 70: 223–87.

Ganopolski, A., and Rahmstorf, S. (2001). 'Rapid changes of glacial climate simulated in a coupled climate model'. *Nature*, 409: 153–8.

Ganopolski A., and Rahmstorf, S. (2002). 'Abrupt glacial climate changes due to stochastic resonance'. *Physical Review Letters*, 88: 038501-1–038501-4.

Ganopolski, A., Petoukhov, V., Rahmstorf, S., Brovkin, V., *et al.* (2001). 'CLIMBER-2: a climate system model of intermediate complexity. Part II: Validation and sensitivity tests'. *Climate Dynamics*, 17 (10): 735–51.

Gornitz, V., and Fung, I. (1994). 'Potential distribution of methane hydrates in the world's oceans'. *Global Biogeochemical Cycles*, 8 (3): 335–47.

Grassberger, P. (1986). 'Do climatic attractors exist?' *Nature*, 323: 609–12.

Grootes, P. M., Stuiver, M., White, J. W. C., Johnson, S., and Jouzel, J. (1993). 'Comparison of oxygen isotope records from the GISP2 and GRIP Greenland ice cores'. *Nature*, 366: 552–4.

Hart, M. H. (1995). 'Atmospheric evolution, the Drake Equation and DNA: sparse life in an infinite universe', in B. Zuckerman and M. H. Hart (eds), *Extraterrestrials—where are they?* Cambridge: Cambridge University Press.

Hasselmann, K. (1963). 'Bispectra of ocean waves', in M. Rosenblatt (ed.), *Time Series Analysis*. New York: Wiley.

Holling, C. S. (1973). 'Resilience and stability of ecological systems'. *Annual Review of Ecological Systems*, 4: 1–23.

ICSU (2001). http://www.icsu.org/DIVERSITAS

IHDP (2001). *International Human Dimensions Programme on Global Environmental Change: Annual Report 2000*. IHDP Secretariat, Walter-Flex-Strasse 3, 53113 Bonn, Germany: IHDP. http://www.uni-bonn.de/ihdp

Jakosky, B. (1998). *The Search for Life on Other Planets*. Cambridge: Cambridge University Press.

Johnson, R. G. (1997). 'Climate control requires a dam at the Strait of Gibraltar'. *EOS—Transactions of the American Geophysical Union*, 78 (277): 280–1.

Johnson, S. D. (1999). 'Markov model studies of the El Niño–Southern Oscillation'. University of Washington, Ph.D. thesis.

Johnson, S. D., Battisti, D. S., and Sarachik, E. S. (2000), 'Empirically derived Markov models and prediction of tropical Pacific sea surface temperature anomalies'. *Journal of Climate*, 13: 3–17.

Jolly, A. (1999). 'The fifth step'. *New Scientist*. 164 (2218): 78–9.

Kabat, P., Claussen, M., Dirmeyer, P. A., Gash, J. H. C., *et al.* (forthcoming), *Vegetation, Water, Humans and the Climate: A New Perspective on an Interactive System*. Berlin: Springer.

Kates, W. R., Clark, W. C., Corell, R., Hall, J. M., *et al.* (2001). 'Sustainability science'. *Science*, 202 (5517): 641–2.

Kirchner, J. W. (1989). 'The Gaia hypothesis: can it be tested?' *Rev. Geophys.*, 27: 223–35.

Kvenvolden, K. A. (1999). 'Potential effects of gas hydrate on human welfare'. *Proceedings of the National Academy of Science of the United States of America*, 98: 3420–6.

Lineweaver, C. H. (2001). 'An estimate of the age distribution of terrestrial planets in the universe: quantifying metallicity as a selection effect'. *Icarus*, 151: 307–13.

Lovelock, J. E. (1995). *The Ages of Gaia: A Biography of our Living Earth*. Oxford: Oxford University Press.

Lovelock, J. E., and Margulis, L. (1974). 'Homeostatic tendencies of the Earth's atmosphere'. *Origins of Life and Evolution of the Biosphere*, 5: 93–103.

Marcy, G. (2001). 'Planet hunter'. *New Scientist*, 169 (2273): 44–7.

Marcy, G. W., and Butler, R. P. (2000). 'Millennium essay: planets orbiting other suns'. *Publications of the Astronomical Society Pacific*, 112: 137–40.

Marcy, G. W., Cochran, W. D., and Mayor, M. (2000). 'Extrasolar planets around main-sequence stars', in V. Mannings, A. Boss, and S. Russel (eds), *Protostars and Planets IV*. Tucson: University of Arizona Press.

Margulis, L., and Schwartz, K. V. (1988). *Five Kingdoms: An Illustrated Guide to the Phyla of Life on Earth*. New York: Freeman.

Menocal, P. B. de, Ortiz, J., Guilderson, T., Adkins, J., *et al.* (2000). 'Abrupt onset and termination of the African humid period: rapid climate response to gradual insolation forcing'. *Quaternary Science Reviews*, 19: 347–61.

Mienert, J., Posewang, J., Baumann, M., *et al.* (1998). 'Gas hydrate along the northeastern Atlantic margin: possible hydrate-bound margin instabilities and possible release of methane', in J. P. Henriet and J. Mienert (eds), *Gas Hydrates: Relevance to World Margin Stability and Climate Change*, Special Publications, 137. London: Geological Society.

Milankovitch, M. (1969). *Canon of Insolation and the Ice Age Problem*. First pub. 1941. Trans. Israel Program for Scientific Translation. Jerusalem: US Department of Commerce and the National Science Foundation.

Morgan, G. (2001). 'Methodological challenges in the assessment of climate change'. Lecture given at the First Sustainability Days Conference held at the Potsdam Institute for Climate Impact Research (PIK), Potsdam, Germany.

Neelin, J. D., Battisti, D. S., Hirst, A., C., Jin, F.-F., *et al.* (1998). 'ENSO theory'. *Journal Geophysical Research*, 103 (C7): 14261–90.

Newton, P. (1999). 'A manual for planetary management'. *Nature*, 400: 399.

Nicolis, C., and Nicolis, G. (1986). 'Is there a climatic attractor?' *Nature*, 311: 529–32.

Norris, R. D., and Röhl, U. (1998). 'Carbon cycling and chronology of climate warming during the Paleocene/Ecocene transition'. *Nature*, 401: 775–8.

Nowotny, H., Scott, P., and Gibbons, M. (2001). *Re-thinking Science: Knowledge and the Public in an Age of Uncertainty*. Cambridge: Polity Press.

Omstedt, A., and Axell, L. (1998). 'Modelling the seasonal, interannual and long-term variations of salinity and temperature in the Baltic proper'. *Tellus*. 50A (5): 637–52.

Penland, C., and Magorian, T. (1993). 'Prediction of Niño 3 sea surface temperatures using linear inverse modeling'. *Journal of Climate*, 6: 1967–76.

Petit, J. R., Jouzel, J., Raynaud, D., Barkov, N. I., *et al.* (1999). 'Climate and atmospheric history of the past 420,000 years from the Vostok ice core, Antarctica'. *Nature*, 399: 429–36.

Petoukhov, V., Ganopolski, A., Brovkin, V., Claussen, M., *et al.* (2000). 'CLIMBER–2: a climate system model of intermediate complexity. Part I: Model description and performance for present climate'. *Climate Dynamics*, 16 (1): 1–17.

Petschel-Held, G., Schellnhuber, H.-J., Bruckner, T., and Tóth, F. (1999). 'The tolerable windows approach: theoretical and methodological foundations'. *Climatic Change*, 41: 303–31.

Pielke Sr., R. A., *et al.* (forthcoming). 'Earth system group report'. Nonlinear Responses to Global Environmental Change: Critical Thresholds and Feedbacks Workshop, IGBP Nonlinear Initiative, Duke University, Durham, NC, 27–9 May 2001.

Rahmstorf, S. (1996). 'On the freshwater forcing and transport of the Atlantic thermohaline circulation'. *Climate Dynamics*, 12: 799–811.

Rahmstorf, S. (1998). 'Influence of the Mediterranean outflow in climate'. *EOS—Transactions of the American Geophysical Union*, 79: 281–2.

Rahmstorf, S. (2000). 'The thermohaline ocean circulation: a system with dangerous thresholds?' *Climatic Change*, 46: 247–56.

Rawls, J. (1999). *A Theory of Justice*. Cambridge, Mass.: Belknap Press.

Rhie, S. H., Bennett, D. P., Fragile, P. C., *et al.* (1998). 'A report from microlensing planet search collaboration: a possible Earth mass planetary system found in MACO–98-BLG–35?' *Bulletin of the American Astronomical Society*, 30 (4): 1415.

Rial, J. A., and Anaclerio, C. A. (2000). 'Understanding nonlinear responses of the climate system to orbital forcing'. *Quaternary Science Reviews*, 19: 1709–22.

Rosenblum, M. G., Pikovsky, A. S., and Kurths, J. (1996). 'Phase synchronization of chaotic oscillators'. *Physical Review Letters*, 76 (11): 1804–7.

Rosnay, J. de (1995). *L'Homme symbiotique*. Paris: Seuil.

Scheffer, M., Carpenter, S., Foley, J. A., Folke, C., and Walker, B. (2001). 'Catastrophic shifts in ecosystems'. *Nature*, 413: 591–6.

Schellnhuber, H.-J. (1998). 'Discourse: Earth system analysis—the concept', in H.-J. Schellnhuber and V. Wenzel (eds), *Earth System Analysis*. Berlin: Springer.

Schellnhuber H.-J. (1999). '"Earth system" analysis and the second Copernican revolution'. *Nature*, 402, suppl. C19–C23.

Schellnhuber, H.-J. (2000). 'The Waikiki principles: rules for a new GAIM'. *IGBP Newsletter*, 41: 3–4.

Schneider, J. (2002). *Extrasolar Planets Catalogue*. http://www.usr.obspm.fr/planets

Schneider, S. H. (1997). 'Integrated assessment modeling of global climate change: transparent rational tool for policy making or opaque screen hiding value-laden assumptions?' *Environmental Modeling Assessment*, 2: 229–49.

Shackley, S., Young, P., Parkinson, S., and Wynne, B. (1998). 'Uncertainty, complexity and concepts of good science in climate change modelling: are GCMs the best tools?' *Climatic Change*, 38: 159–205.

Smith, J. B., Schellnhuber, H.-J., and Mirza, M. M. Q. (2001). 'Vulnerability to climate change and reasons for concern: a synthesis', in J. J. McCarthy, O. F. Canziani, N. A. Leary, D. J. Dokken, K. S. White (eds), *Climate Change 2001: Impacts, Adaptation and Vulnerability. IPCC Third Assessment Report.* Cambridge: Cambridge University Press.

Steffen, W., and Tyson, P. (eds) (2001). *Global Change and the Earth System: A Planet under Pressure*, IGBP Science Paper, no. 4. Stockholm: IGBP.

Stommel, H. (1961). 'Thermohaline convection with two stable regimes of flow'. *Tellus*, 13: 224–30.

Terzan, Y., and Bilson, E. (eds) (1997). *Carl Sagan's Universe.* Cambridge: Cambridge University Press.

Titz, S., Kuhlbrodt, T., and Feudel, U. (forthcoming). 'Homoclinic bifurcation in an ocean circulation box model'. *International Journal of Bifurcation and Chaos.*

Tóth, F. L., Bruckner, T., Fuessel, H.-M., Leimbach, M., *et al.* (forthcoming). 'Exploring climate policy fields in an inverse integrated assessment framework'.

Tziperman, E., and Scher, H. (1997), 'Controlling spatiotemporal chaos in a realistic El Niño prediction model'. *Physical Review Letters*, 79: 1034–7.

Volk, T. (1997). *Gaia's Body.* Berlin: Springer.

Walker, G. T., and Bliss, E. W. (1932). 'World Weather V'. *Memoirs of the Royal Meteorological Society*, 4: 53–84.

Wallace, J. M., and Gutzler, D. S. (1981). 'Teleconnections in the geopotential height field during the northern hemisphere winter'. *Monthly Weather Review*, 109: 784–812.

Wang, G., and Eltahir, E. A. B. (2000). 'Biosphere–atmosphere interaction over West Africa, 2: Multiple climate equilibria'. *Quarterly Journal of the Royal Meteorological Society*, 126: 1261–80.

WBGU (1995). Wissenschaftlicher Beirat der Bundesregierung Globale Umweltveränderungen, *Szenario zur Ableitung globaler CO_2-Reduktionsziele und Umsetzungsstrategien.* Bremerhaven: WBGU.

WCRP (2001). *World Climate Research Programme.* http://www.wmo.ch/web/wcrp/about.htm

Webster, P. J., Magaña, V. O., Palmer, T. N., Shukla, J., *et al.* (1998). 'Monsoons: processes, predictability, and the prospects for prediction'. *Journal Geophysical Research*, 103 (C7): 14451–510.

Zebiak, S. E., and Cane, M. A. (1987). 'A model El Niño–Southern Oscillation'. *Monthly Weather Review*, 115: 2262–78.

2

Risks of Conflict: Resource and Population Pressures

Crispin Tickell

INTRODUCTION

CONCERN about resources depletion and population pressure has risen and fallen over the last few years. It arouses strong feelings, not least because it is hard to define—or even to identify with certainty—and still harder to cope with. The implications go far and wide. Many people are reluctant to take them seriously.

Wars, conflicts, and the use of force are the endemic conditions of humanity. Few like it. Almost all condone it. The reasons are almost as various as people themselves: ideology, fear, glory, greed, habit, pressure on resources, and the compulsions of power all play a part. Fights between our near cousins the chimpanzees may be painful for them but they leave little trace afterwards. Only with the enormous growth in the human population and its ever increasing demands has war affected our physical surroundings, and in the last century the good health of the planetary environment as a whole.

The two world wars in the last century showed how dangerous wars between industrial states had become. It is probably no coincidence that there has been none since then. As the United Nations Secretary-General remarked in his Millennial Report in 2000, 'the majority of wars today are wars among the poor'. Violence within societies has also increased. Of the twenty-seven armed conflicts that took place in 1999, all but two were within national boundaries. It is one of the products of the current decline in the power and status of the nation state. Power is moving upwards to international institutions (however ill equipped to receive it) to cope with problems outside the abilities of any state; it is moving downwards to regional, ethnic, and local communities anxious to recover identity and meaning; and it is moving sideways between citizens worldwide through the marvels of radio, television, the internet, and other electronic means of communication.

In these changing circumstances the world looks a messier place than I have known it during my thirty-six years as a diplomat. The pressures on human society are increasing relentlessly. A recent book on the environmental history of the twentieth century was well entitled *Something New Under the Sun* (McNeill 2000). Up to now there has been a precedent for most things: population explosions of particular plants or animals; periodical extinctions; changes in soil fertility; rapid global cooling and rapid global warming; even impacts of objects from outer space. But in the history of life there are few, if any, precedents for the enormous impact on the earth of one animal species: our own. It can be seen as a case of malignant maladaptation in which a species, like infected tissue in the organism of life, multiplies out of control, affecting everything else. In terms of factors of increase within the last century, the human population rose by four, air pollution by around five, water use by nine, sulphur emissions by thirteen, energy use by sixteen, carbon dioxide emission by seventeen, marine fish catch by thirty-five, and industrial output by forty (McNeill 2000).

REASONS FOR CONFLICT

Most of the implications of this growth remain unrecognized. Our lives are too short to grasp them; but nearly all have potentialities for violence. With human communities everywhere under strain, there are five main drivers for change in the human condition, each associated with the others, and all pointing towards risks of social breakdown, which in turn could lead to violence in one form or another. They are the rate of human population increase; change in the condition of the land surface; water; damage to natural ecosystems; and change in the chemistry of the atmosphere.

First comes the rate of human population increase. At the end of the last Ice Age some 12,000 years ago the human population was probably around 10 million. With the introduction of agriculture, urbanization, and complex social organization, numbers rose steeply. At the time of Thomas Malthus at the end of the eighteenth century, when industrialization was just beginning, the population was around 1 billion. By 1930 it had risen to 2 billion. Today it is over 6 billion, and according to most predictions it will rise to between 8 and 10 billion by the middle of this century. Between the Rio Conference of 1992 and 2000, some 450 million new people came to inhabit the earth. This is more than the total population at the time of the

Roman and Han empires some 2,000 years before. If the increase had been in elephants, swallows, sharks, or cockroaches, we would have been scared silly; but, as it is ourselves, we shrug our shoulders as if it were the most normal thing in the world.

Within the figures there are some clear trends. One is towards urbanization. More than half the human species now lives in cities, and that proportion is increasing every year. Such cities absorb more and more resources from the area around them and, seen from space, would look like some sort of melanoma on the skin of the earth. They can be compared to living organisms with inputs and outputs, and lives and deaths. Many in poor countries have become virtually unmanageable, and, as in the Middle Ages, create dangerous problems for human health.

A second trend is towards differential rates of population increase. The curves can be interpreted in different ways, but the age profiles tell a consistent story. With improvements in health and comfort, the population in industrial countries is roughly stable and getting older; and in poorer countries it is unstable and getting younger. Even if rates of increase are at last declining in some poorer parts of the world, the aggregate rate of increase is still upwards. No wonder that many believe that these numbers exceed the carrying capacity of the earth for our particular species.

A third trend is a widening gap between rich and poor. Current projections suggest that the number of people living in absolute poverty is likely to rise from 1.3 billion to 1.6 billion in 2010. In his Millennium Report in 2000 the UN Secretary-General used a telling image of the global village. Supposing that village contained some 1,000 individuals, 150 would live in affluent areas, some 780 in poor areas, and some seventy or so between the two. Just 200 would dispose of 86 per cent of the village's wealth; some 220 (two-thirds of them women) would be illiterate; of the 390 inhabitants under the age of 20, three-quarters would live in the poorer areas, and many of them would be looking for work. Life expectancy in the rich areas would be around 78 years, and in the poorer 52 years. The poor areas would be much more prone to disease, and would lack safe water, sanitation, health care, good housing, education, and useful employment. Differences in consumption would be enormous. So would understanding and practice of information technology. Taken together these elements suggest a pretty unstable village life. Indeed they make a combustible mixture.

The second driver is change in the condition of the land. More people need more space and more resources. Soil degradation is estimated to affect some 10 per cent of the current world agricultural area. Although more and more land, whatever its quality, is used for human purposes, increase in food supplies has not kept pace with increase in population. Today many of the problems are of distribution. But even countries generating food surpluses can see limits ahead. Application of biotechnology, itself with some dubious aspects, could never hope to meet likely shortfalls.

In the meantime industrial contamination of various kinds has greatly increased. To run our complex societies, we need copious amounts of energy, at present overwhelmingly derived from the world's dwindling resources of fossil fuels laid down hundreds of millions of years ago. We also have to deal with the mounting problems of waste disposal, including the toxic products of industry.

The third driver is water. No resource is in greater demand than fresh water. At present such demand doubles every twenty-one years, and seems to be accelerating. Yet supply in a world of 6 billion people is the same as it has been for thousands of years. Falling water-tables are widespread and cause serious problems because they lead both to water shortages and in coastal areas to salt intrusion. Already the Global Environment Outlook 2000 (United Nations Environment Programme 1999) refers to a global water crisis. In the future we must expect water to be a still greater cause of dissension and conflict than in the past.

At the same time there has been increasing pollution of water, both fresh and salt. Some rivers, for example the Yellow River in China and the River Jordan, have dwindled into noxious streams. It can be no surprise that coastal areas are at particular risk from materials brought down by rivers to the sea. There is an increasing incidence of toxins produced by blue-green algae along rivers and coasts, and in the deep oceans fish stocks are rapidly declining. Indeed fishing fleets are 40 per cent larger than fishing stocks can sustain.

The fourth driver is damage to the natural ecosystems of which humans are a small but immodest part. Human activities are causing extinction of other species at around a thousand times the natural rate, while the replacement rate is as always slow and capricious. Too easily we forget human dependence on other species. Reduction of biodiversity affects food supplies (already heavily dependent on a few genetic strains) and medicine (heavily dependent on limited plant and animal sources). But more important are the ecological benefits: we rely on forests and vegetation to produce soil, to hold it together, and to regulate supplies by preserving catchment basins, recharging groundwater and buffering extreme conditions; we rely on soils to be fertile and break down pollutants; and we rely on nutrients for recycling and disposal of waste. There is no conceivable substitute for such natural services. At present there is a general impoverishment of the biosphere. According to the Living Planet Index (WWF International 2000), the state of the Earth's natural ecosystems has declined by about a third in the last thirty years, while the ecological pressure of humanity has increased by about a half during the same period. These processes are accelerating.

The fifth driver is change in the chemistry of the atmosphere. Acidification from industry has affected wide areas downwind. Depletion of the atmospheric ozone layer is permitting more ultra-violet radiation to reach the surface of the earth, with so far unmeasured effects on organisms unadapted to it. Greenhouse gases are increasing at a rate which could change average world temperature, with big resulting variations in climate and local weather as well as sea levels. The latest reports from Working Groups 1 and 2 of the Intergovernmental Panel on Climate Change well illustrate the scope of the problem. By any reckoning their conclusions are alarming, and the world has yet to take proper account of them. Here the British Prime Minister, following his two predecessors in office, gave a real lead in his speech of 6 March 2001.[1] The debate about climate has shifted from whether change is happening at all to the degree of change and how it will affect different parts of the world. The recent El Niño phenomenon has well illustrated what vast changes can be caused by small perturbations in the climate system.

The implications for security are obvious. So long as human numbers were relatively small, changes in climate, as over the last 10,000 years, could be accommodated by movements of people. But in a crowded world, where people are up against natural limits, big movements of population, as after the last Ice Age, are not practical politics. No wonder that climate change has often been identified as the single biggest threat to world stability.

Its manifestations in the form of drought, floods, or sea-level rise could lead to any number of conflicts. But it is rather the combination of the circumstances I have described which is most important. In 1972 a prophetic book was published entitled *The Limits to Growth* (Meadows *et al.* 1972). Many people poured scorn on it as an extrapolation of existing trends. This was somewhat unfair as the authors were simply trying to warn about what might happen if such trends continued.

Twenty years later the same authors published *Beyond the Limits* (Meadows *et al.* 1992). In it they showed that on current models, with expanding human population and continuing economic growth, the world economy as it is now known will eventually be unable to function: not just for the obvious reasons, important as they are, but because governments and communities would simply be unable to cope. There would be a creeping contagion of breakdown in different countries at different times. The problem of what has been called state failure is very real. There have been cases in which state bankruptcy has led society to collapse like a subsiding sandhill, with the inhabitants of cities returning to an unwelcoming reception in their

---

[1] *Environment: The Next Step.* Speech by Tony Blair at Chatham House, London, 6 Mar. 2001.

countryside of origin, or even less welcoming neighbour countries. This has happened already, for example in West Africa, Haiti, and parts of Asia, including Indonesia.

PROSPECTS FOR CONFLICT

So if the prospects for conflict are limitless, what are the likely triggers? How can we be more precise? Nothing is more difficult, not least because circumstances are so different, and effects do not directly follow causes. In poor countries depletion or degradation of natural resources, for example water, land for crops, fuel wood, and fish, have a multiplicity of social consequences. They create poverty, promote inequity, aggravate tensions within communities, and weaken institutions. In industrial countries technology can sometimes hold such problems at bay, at least for a while. At present they are dependent on access to fossil fuels for their energy supplies, and most of these, including some offshore, are inconveniently situated outside their control. They are generally more vulnerable than they suppose, as was brought out recently in a paper entitled *The Future Strategic Context for Defence* (Ministry of Defence 2001). Many of the same points appear in a CIA paper entitled *Global Trends 2015* (US National Intelligence Council 2000), although here the perennial US spirit of optimism permeates the analysis.

The triggers for conflict within and between countries and communities can come almost anywhere, and once conflicts start, they are likely to persist and to spread. Here are some examples. Water shortage, increased by depletion of aquifers, is one of the underlying sources of tension between Israelis and Palestinians. At present illegal Israeli settlers in the Occupied Territories take up much more water than the Palestinian inhabitants. Destruction of forest cover and degradation of cropland, combined with seizures of communal lands, in the Mexican state of Chiapas are the principal source of the recent Zapatista uprising against the central government. Comparable problems exist in many parts of Africa, in particular Ethiopia and Rwanda. Soil erosion in Nepal, Haiti, the slopes of the Andes, and parts of China have had profoundly destabilizing effects with consequences that governments find increasingly difficult to manage.

So far conflicts between nation states have been rare, with the exception of those arising from oil. This may not continue. Water is one prime threat. For example, any attempt by the Ethiopian government to divert or diminish the flow of water from Lake Tana into the Blue Nile could be regarded by Egypt or the Sudan as a *casus belli*. The export of pollution of any kind across frontiers could be the same. What states do to the environment within their boundaries is no longer for them alone. Some might even consider deliberate damage to the environment as a tool of war. This has happened often enough in the past. One manifestation is scorching of the earth. Certainly the effects of war on the environment, ranging from Hiroshima to Kosovo, can be catastrophic. When I raised these possibilities in the UN Security Council at the end of the 1980s, many did not want to hear me. I fear they may have to listen in the future.

REFUGEES

Another prime threat arises from refugees. They are a consequence as well as a cause of destabilization. Refugees fall into three broad categories: political refugees (those covered by the United Nations High Commission for Refugees); economic migrants (those who move from poorer to apparently richer parts of the world); and environmental refugees (those driven out of their homes by changing environmental conditions). Of course there is some blurring between the three groups. Yet there has been a reluctance to recognize environmental refugees as such. All categories have greatly increased in number over the last quarter-century. In January 2000 there were over 22 million political and other conventionally defined refugees (United Nations High Commission for Refugees 2000). Environmental refugees, as they have no legal status, are more difficult to quantify, but a 1995 estimate put them at 25 million, with particularly large numbers in Africa south of the Sahara (Myers and Kent 1995). In the circumstances I have described, the total number could greatly increase during this century.

It is worth looking in more detail at one of the issues raised by refugees. At present a heavy concentration of people is living in low-lying coastal areas along the world's great river systems. Nearly one-third of humanity lives within 60 kilometres of a coastline. A rise in mean sea level of only 25 centimetres would have substantial effects, and the predictions are for much more in the next 100 years. The industrial countries might be able to construct new sea defences to protect vulnerable areas, but even they would have difficulty coping with high tides and storm surges of a kind likely to be more common.

For most poor countries such defences would be out of the question. Many of those living and working in, for example, the delta areas of the Nile, the Ganges, the Yangtze, and the Zambezi would be forced out of their homes and livelihood. Such islands as the Maldives in the Indian Ocean, and Kiribati, Tuvalu, and the Marshall Islands in the Pacific, would soon become uninhabitable. Bangladesh with its population of well over 100 million, and Egypt with its population of almost as much, would be particularly affected. A rise in sea level beyond half a metre would have more drastic results. The world would look a different place.

What then could be the scale of the human problem in 2010, 2020, or 2040? Even allowing for piecemeal, gradual responses year by year to the imperatives of change, it would be very large. Plucking a figure from the air, if only 1 per cent (a very low estimate) of a world population of 8 billion in 2040 were affected, that would still mean some 80 million refugees of all kinds; and 5 per cent (again a low estimate) would produce 400 million. Even 80 million represents a problem of an order of magnitude which no one has yet had to face.

Nor is it all. Refugees create their own problems. In some cases they can return to their countries of origin. In others they can be resettled. Economic migrants or environmental refugees fall into another category whether they stay in their own country or cross into another. At present they are mostly a phenomenon in poor countries. Shelter, food, and medical care are hard to find. There is little prospect of return. They often come from another environment (for example, highland Ethiopians are forced down to the plains), and bring with them alien customs, religious practices, eating habits, agricultural methods, and—not least—diseases, with susceptibility to local pathogens.

Most have great difficulty in adjusting themselves to new circumstances. Like normal refugees, they mostly depend on charity. Resettlement is never easy, and full assimilation is rare and inevitably slow. In any numbers they tend to spread their poverty around them, and to compound the problem from which they first tried to escape. In a world of rapid change they would constitute only one of myriad animal species trying to cope with disruption of their way of life.

Within a country refugees would represent a dangerous element in what would anyway be increasing difficulties of social and economic management. Some governments can cope, some evidently cannot. But few outside the industrial world have the structure or resources to manage a continuing crisis. Such secondary effects as disorder, terrorism, economic breakdown, disease, or bankruptcy could become endemic. We are all too familiar with them already.

Between countries and regions there would be still greater difficulties. To a greater or lesser extent all would be suffering, and undergoing adjustment. Thus willingness to help others would be limited, the more so if such help threatened to put at risk the adjustment process at home. In times of trouble the pressure of recognizable aliens is liable to ignite popular resentment with the speed of a brushfire.

In industrial countries many feel, rightly or wrongly, that there is an absolute limit to the number of people from other countries and cultures which can be absorbed without damaging social cohesion and national identity. Some of them certainly welcome migrants with particular qualities and skills. Countries with ageing populations could find a well-controlled infusion of the young an advantage. But general resistance to refugees has become popular politics. They can even bedevil foreign policy. Often Albanian, Kurdish, Tamil, Ethiopian, or Eritrean refugees want to use their safety abroad to play politics at home. This can be profoundly embarrassing to the host country. Any aggravation of the refugee problem would only strengthen such resistance.

But even if some people and governments wished to seal themselves off from the rest of the world, like Americans in suburban enclaves, they could not do so. In no country or city can the rich fortify themselves for long against the poor. All form part of an increasingly interdependent society. Land frontiers can always be penetrated. The movement northwards of Mexicans and other Latin Americans across the long southern frontier of the United States has so far proved irresistible, and every year parts of the United States develop more Hispanic characteristics. Nor are short sea crossings a real barrier. Desperation could push Africans into Europe, Chinese into the relatively empty parts of Russia, and Indonesians into northern Australia. Sheer numbers could swamp most efforts at control. Yet the refugee issue remains strangely low on the world's long-term agenda. The issue may be too hot to handle now. But it will surely be still hotter in the future.

CONCLUSION

Looking ahead at the prospects for conflict, we seem to be in for a bumpy ride. Violence within and between communities and between nation states could well increase. The precedents are all around us. It would be naive to expect otherwise, and we must be prepared for it. Efforts have been made to work out sets of indicators to show when and where problems are likely to reach danger point. The trouble is that almost every one is special with its own characteristics. No one knows which way the cat will jump or if it will jump at all. Water scarcity could lead to war, or it could lead to cooperation.

In some cases things may have to get worse before they can get better. There is nothing like a catastrophe to concentrate the mind. Certainly we need to think differently about our place in the natural world, the numbers that the world can support, the ways in which we are using or abusing its resources, the philosophies of exploitation and consumerism which guide our society, even the methods by which we measure wealth and well-being.

As animals we are both tough and adaptable. But our toughness and adaptability might be tested beyond endurance. For many, that is when the fighting begins. We have grown, lived, and flourished as elements in specific social and environmental surroundings. Those surroundings or ecosystems could be so damaged that they could fall apart, as they often have in the past, and be replaced by different ones. We should remember the fate of the thirty or so societies which have preceded our own, and left no more than traces behind them.

There are no magic solutions. Obviously we need to be able to contain violence, and to have enforceable rules to govern relations not only between states but also between governments and their citizens. There is a panoply of international institutions, conventions, and declarations for the purpose. But perhaps more important is to go for the underlying causes of conflict and try to diminish or mitigate the consequences. Old Adam and old Eve are with us—competitive, docile, various, peaceful, violent, creative, and wasteful—now as in the future.

It is a poor lookout. In his life of Samuel Johnson, James Boswell referred to the would-be philosopher Oliver Edwards, who once said of the gloom that was expected of him: 'I don't know how, cheerfulness was always breaking in.' Let us leave it that way.

REFERENCES

McNeill, J. (2000). *Something New under the Sun: An Environmental History of the 20th Century*. London: Allen Lane Penguin Press.

Meadows, Donella H., Meadows, Dennis L., and Randers, J. (1972). *The Limits to Growth*. New York: New York Books.

Meadows, Donella H., Meadows, Dennis L., and Randers, J. (1992). *Beyond the Limits: Global Collapse or a Sustainable Future*. London: Earthscan.

Ministry of Defence (2001). *The Future Strategic Context for Defence*. London: Ministry of Defence.

Myers, Norman, and Kent, Jennifer (1995). *Environmental Exodus*. Washington, DC: Climate Institute.

Summaries of Reports by Working Groups 1 and 2 as part of the Third Assessment of the Intergovernmental Panel on Climate Change (2001). Cambridge: Cambridge University Press.

United Nations Environment Programme (1999). *Global Environment Outlook (GEO) 2000*. London: Earthscan.

United Nations High Commission for Refugees (2000). *Annual Report 2000*. Geneva: UNHCR.

United Nations Secretary General's Millennium Report (2000). *We the Peoples: The Role of the United Nations in the 21st Century*: New York: United Nations.

US National Intelligence Council (2000). *Global Trends 2015*. Washington, DC: US Government Printing Office.

WWF International (2000). *Living Planet Report 2000*. Gland: WWF International.

3

Valuing the Earth: Reintegrating the Study of Humans and the Rest of Nature

Robert Costanza

INTRODUCTION

PRACTICAL problem-solving in complex, human-dominated ecosystems requires the integration of three elements: (1) active and ongoing envisioning of both how the world works and how we would like the world to be; (2) systematic analysis appropriate to and consistent with the vision; and (3) implementation appropriate to the vision. Scientists generally focus on only the second of these steps, but integrating all three is essential to both good science and effective management. 'Subjective' values enter in the 'vision' element, both in terms of the formation of broad social goals and in the creation of a 'pre-analytic vision' which necessarily precedes any form of scientific analysis.

Because of this need for vision, completely 'objective' scientific analysis is impossible. In the words of Joseph Schumpeter (1954: 41):

In practice we all start our own research from the work of our predecessors, that is, we hardly ever start from scratch. But suppose we did start from scratch, what are the steps we should have to take? Obviously, in order to be able to posit to ourselves any problems at all, we should first have to visualise a distinct set of coherent phenomena as a worthwhile object of our analytic effort. In other words, analytic effort is of necessity preceded by a preanalytic cognitive act that supplies the raw material for the analytic effort. In this book, this preanalytic cognitive act will be called Vision. It is interesting to note that vision of this kind not only must precede historically the emergence of analytic effort in any field, but also may reenter the history of every established science each time somebody teaches us to *see* things in a light of which the source is not to be found in the facts, methods, and results of the pre-existing state of the science.

Nevertheless, it is possible to separate the process into the more subjective (or normative) envisioning component, and the more systematic, less subjective analysis component (which is based on the vision). 'Good science' can do no better than to be clear about its underlying pre-analytic vision, and to do analysis that is consistent with that vision.

A CHANGING VISION OF SCIENCE

The task would be simpler if the vision of science were static and unchanging. But as the quote from Schumpeter above makes clear, this vision is itself changing and evolving as we learn more. This does not invalidate science, as some deconstructionists would have it. Quite the contrary, by being explicit about its underlying pre-analytic vision, science can enhance its honesty and thereby its credibility. This credibility is a result of honest exposure and discussion of the underlying process and its inherent subjective elements, and a constant pragmatic testing of the results against real world problems, rather than by appeal to a non-existent objectivity.

The pre-analytic vision of science is changing from the logical positivist view (which holds that science can discover ultimate 'truth' by falsification of hypothesis) to a more pragmatic view that recognizes that we do not have access to any ultimate, universal truths, but only to useful abstract representations (models) of small parts of the world. Science, in both the logical positivist and in this new pragmatic modelling vision, works by building models and testing them. However, the new vision recognizes that the tests are rarely, if ever, conclusive (especially in the life sciences and the social sciences); the models can only apply to a limited part of the real world; and the ultimate goal is therefore not 'truth' but quality and utility. In the words of William Deming, 'All models are wrong, but some models are useful' (McCoy 1994).

The primary goal of science is then the creation of useful models whose utility and quality can be tested against real-world applications. The criteria by which one judges the utility and quality of models are themselves social constructs which evolve over time. There is, however, fairly broad and consistent consensus in the peer community of scientists about what these criteria are. They include (1) testability, (2) repeatability, (3) predictability, and (4) elegance (i.e. Occam's razor—the model should be as simple as possible—but no simpler!). Because of the nature of real-world problems, there are many applications for which some of these criteria are difficult or impossible to apply. These applications may nevertheless still be judged as 'good science'. For example, some purely theoretical models are not directly testable, but they may provide a fertile ground for thought

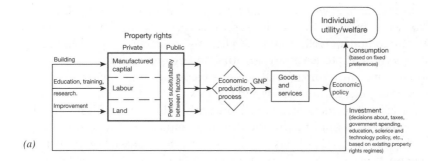

(a)

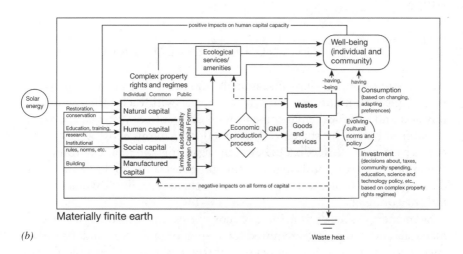

(b)

Fig. 3.1. (*a*) Conventional 'empty world' model of the economy. (*b*) Expanded 'full world' model of the ecological economics model.

Social capital is a recent concept that includes the web of interpersonal connections, institutional arrangements, rules, and norms that allow individual human interactions to occur (Berkes and Folke 1994). Property rights regimes in this model are complex and flexible, spanning the range from individual to common to public property. Natural capital captures solar energy and behaves as an autonomous complex system and (unlike the conventional model) this model conforms to the basic laws of thermodynamics requiring the conservation of mass and energy, and the entropic degradation of all complex structures. Natural capital contributes to the production of marketed economic goods and services, which affect human welfare. It also produces ecological services and amenities that directly contribute to human welfare without ever passing through markets. There is also waste production by the economic process, which contributes negatively to human welfare and has a negative impact on capital and ecological services. Preferences are adapting and changing, but basic human needs are constant. Human welfare is a function of much more than the consumption of economic goods and services.

These visions of the world are significantly different. As Ekins (1992: 151) points out: 'It must be stressed that the complexities and feedbacks of model b [see Fig. 3.1(*b*) of this paper] are not simply glosses on model a's simpler portrayal of reality. They fundamentally alter the perceived nature of that reality and in ignoring them conventional analysis produces serious errors ...'.

ECOSYSTEMS AND ECOSYSTEM SERVICES

An *ecosystem* consists of plants, animals, and micro-organisms which live in biological communities and which interact with each other, with the physical and chemical environment, with adjacent ecosystems, and with the atmosphere. The structure and functioning of an ecosystem is sustained by synergistic feedbacks between organisms and their environment. For example, the physical environment puts constraints on the growth and development of biological subsystems, which, in turn, modify their physical environment.

Solar energy is the driving force of ecosystems, enabling the cyclic use of materials and compounds required for system organization and maintenance. Ecosystems capture solar energy through photosynthesis by plants. This is necessary for the conversion, cycling, and transfer to other systems of materials and critical chemicals that affect growth and production, i.e. biogeochemical cycling. Energy flow and biogeochemical cycling set an upper limit on the quantity and number of organisms, and on the number of trophic levels that can exist in an ecosystem (Odum 1989).

Holling (1986) has described ecosystem behaviour as the dynamic sequential interaction between four basic system functions: exploitation, conservation, release, and reorganization. The first two are similar to traditional ecological succession. *Exploitation* is represented by those ecosystem processes that are responsible for rapid colonization of disturbed ecosystems during which organisms capture easily accessible resources. *Conservation* occurs when the slow resource accumulation builds and stores increasingly complex structures. Connectedness and stability increase during the slow sequence from exploitation to conservation and a 'capital' of biomass is slowly accumulated. *Release* or *creative destruction* takes place when the conservation phase has built elaborate and tightly bound structures that have become 'over-connected', so that a rapid change is triggered. The system has become *brittle*. The stored capital is then suddenly released and the tight organization is lost. The abrupt destruction is created internally but caused by an external disturbance such as fire, disease, or grazing pressure. This process of change both destroys and releases opportunity for the fourth stage, *reorganization*, where released materials are mobilized to become available for the next exploitive phase.

The stability and productivity of the system is determined by the slow exploitation and conservation sequence. *Resilience*, that is, the system's capacity to recover after disturbance—its capacity to absorb stress—is determined by the effectiveness of the last two system functions. The self-organizing ability of the system, or, more particularly, the resilience of that self-organization, determines its capacity to respond to the stresses and shocks imposed by predation or pollution from external sources.

Some natural disturbances, such as fire, wind, and herbivores, are an inherent part of the internal dynamics of ecosystems, and in many cases set the timing of successional cycles (Holling *et al.* 1995). Natural perturbations are parts of ecosystem development and evolution, and seem to be crucial for ecosystem resilience and integrity. If they are not allowed to enter the ecosystem, it will become even more brittle and thereby even larger perturbations will be invited with the risk of massive and widespread destruction. For example, small fires in a forest ecosystem release nutrients stored in the trees and support a spurt of new growth without destroying all the old growth. Subsystems in the forest are affected but the forest remains. If small fires are blocked out from a forest ecosystem, forest biomass will build up to high levels and when the fire does come it will wipe out the whole forest. Such events may flip the system to a totally new state that will not generate the same level of ecological functions and services as before (Holling *et al.* 1995). These sorts of flips may occur in many ecosystems. For example, savannah ecosystems (Perrings and Walker 1995), coral reef systems (Knowlton 1992), and shallow lakes (Scheffer *et al.* 1993) can all exhibit this kind of behaviour. The flip from one state to another is often induced by human activity; for example, cattle-ranching in savannah systems can lead to completely different grass species assemblages and nutrient enrichment. Physical disturbance around coral reefs can lead to replacement with algae-dominated systems, and nutrient additions can lead to eutrophication of lakes.

Natural ecosystems, including human-dominated systems, have been called 'complex adaptive systems'. Because these systems are evolutionary rather than mechanistic, they exhibit a limited degree of predictability.

Understanding the problems and constraints which these evolutionary dynamics pose for ecosystems is a key component in managing them sustainably (Costanza *et al.* 1993).

Ecological systems play a fundamental role in supporting life on earth at all hierarchical scales. They form the life support system without which economic activity would not be possible. They are essential in global material cycles such as the carbon and water cycles. Ecosystems produce renewable resources and ecological services. For example, a fish in the sea is produced by several other 'ecological sectors' in the food web of the sea. The fish is a part of the ecological system in which it is produced, and the interactions that produce and sustain the fish are inherently complex.

Ecological services are those ecosystem functions that are currently perceived to support and protect human activities or affect human well-being (Daily 1997; Costanza *et al.* 1997a). They include maintenance of the composition of the atmosphere, amelioration and stability of climate, flood controls and drinking water supply, waste assimilation, recycling of nutrients, generation of soils, pollination of crops, provision of food, maintenance of species and a vast genetic library, and also maintenance of the scenery of the landscape, recreational sites, aesthetic and amenity values (Table 3.1) (Ehrlich and Mooney 1983; Ehrlich and Ehrlich 1992; de Groot 1992; Costanza *et al.* 1997a). Biodiversity at genetic, species, population, and ecosystem levels all contribute in maintaining these functions and services. Cairns and Pratt (1995) argue that if a society were highly environmentally literate, it would probably accept the assertion that most, if not all, ecosystem functions are, in the long term, beneficial to society.

Ecosystem services are seldom reflected in resource prices or taken into account by existing institutions in industrial societies. Many current societies employ social norms and rules that (1) bank on future technological fixes and assume that it is possible to find technical substitutes for the loss of ecosystem goods and services; (2) use narrow indicators of welfare; and (3) employ world-views that alienate people from their dependence on healthy ecosystems. But as the scale of human activity continues to increase, environmental damage begins to occur not only in local ecosystems, but also regionally and globally. Humanity now faces a novel situation of jointly determined ecological and economic systems. This means that as economies grow relative to their life-supporting ecosystems, the dynamics of both become more tightly connected. In addition, the joint system dynamics can become increasingly discontinuous the closer the economic systems get to the carrying capacity of ecosystems (Costanza *et al.* 1993; Perrings *et al.* 1995).

The support capacity of ecosystems in producing renewable resources and ecological services has only recently begun to receive attention, despite

TABLE 3.1. *Ecosystem services and functions*

Ecosystem service[a]	Ecosystem functions	Examples
Gas regulation	Regulation of atmospheric chemical composition	CO_2–O_2 balance, O_3 for UVB protection, and SO_x levels
Climate regulation	Regulation of global temperature, precipitation, and other biologically mediated climatic processes at global or local levels	Greenhouse gas regulation, precipitation, cloud formation
Disturbance regulation	Capacitance, damping, and integrity of ecosystem response to environmental fluctuations	Storm protection, flood control, drought recovery, and other aspects of habitat response to environmental variability mainly controlled by vegetation structure
Water regulation	Regulation of hydrological flows	Provisioning of water for agricultural (e.g. irrigation) or industrial (e.g. milling) processes or transportation
Water supply	Storage and retention of water	Provisioning of water by watersheds, reservoirs, and aquifers
Erosion control and sediment retention	Retention of soil within an ecosystem	Prevention of loss of soil by wind, run-off, or other removal processes, storage of silt in lakes and wetlands
Soil formation	Soil formation processes	Weathering of rock and the accumulation of organic material
Nutrient cycling	Storage, internal cycling, processing, and acquisition of nutrients	Nitrogen fixation, N, P, and other elemental or nutrient cycles
Waste treatment	Recovery of mobile nutrients and removal or breakdown of excess or xenic nutrients and compounds	Waste treatment, pollution control, and detoxification
Pollination	Movement of floral gametes	Provisioning of pollinators for the reproduction of plant populations

TABLE 3.1. (*cont.*)

Ecosystem service[a]	Ecosystem functions	Examples
Biological control	Trophic–dynamic regulations of populations	Keystone predator control of prey species, reduction of herbivory by top predators
Refugia	Habitat for resident and transient populations	Nurseries, habitat for migratory species, regional habitats for locally harvested species, or over-wintering grounds
Food production	That portion of gross primary production extractable as food	Production of fish, game, crops, nuts, fruits by hunting, gathering, subsistence farming, or fishing
Raw materials	That portion of gross primary production extractable as raw materials	The production of lumber, fuel, or fodder
Genetic resources	Sources of unique biological materials and products	Medicine, products for materials science, genes for resistance to plant pathogens and crop pests, ornamental species (pets and horticultural varieties of plants)
Recreation	Providing opportunities for recreational activities	Eco-tourism, sport fishing, and other outdoor recreational activities
Cultural	Providing opportunities for non-commercial uses	Aesthetic, artistic, educational, spiritual, and/or scientific values of ecosystems

[a] Ecosystem 'goods' are included along with ecosystem services.
Source: Costanza *et al.* (1997*a*).

the fact that this 'factor of production' has always been a prerequisite for economic development. In the long run a healthy economy can only exist in symbiosis with a healthy ecology. The two are so interdependent that isolating them for academic purposes has led to distortions and poor management.

A preliminary assessment of the value of ecosystem services at the global scale (Costanza *et al.* 1997*a*) indicated that they provide a significant portion of the total contribution to human welfare on this planet. This study estimated the annual value of these services (in 1994 $US) at $16–54 trillion, with an estimated average of $33 trillion (which is significantly larger than global GNP). Because of the nature of the uncertainties in this estimate, it is almost certainly an underestimate. Coastal environments, including estuaries, coastal wetlands, beds of sea grass and algae, coral reefs, and continental shelves were estimated to be of disproportionately high value in this study. They cover only 6.3 per cent of the world's surface, but are responsible for 43 per cent of the estimated value of the world's ecosystem services. These environments are particularly valuable in regulating the cycling of nutrients that control the productivity of plants on land and in the sea.

DEFINING AND PREDICTING SUSTAINABILITY IN ECOLOGICAL TERMS

Defining sustainability is actually quite straightforward (Costanza and Patten 1995): 'A sustainable system is one which survives or persists.' Biologically, this means avoiding extinction and living to survive and reproduce. Economically, it means avoiding major disruptions and collapses, hedging against instabilities and discontinuities. Sustainability, at its base, always concerns temporality and, in particular, longevity.

The problem with the above definition is that, like fitness in evolutionary biology, determinations can only be made *after the fact*. An organism alive right now is fit to the extent that its progeny survive and contribute to the gene pool of future generations. The assessment of fitness today must wait until tomorrow. The assessment of sustainability must also wait until after the fact.

What often pass as *definitions* of sustainability are therefore usually *predictions* of actions taken today that one hopes will lead to sustainability. For example, keeping harvest rates of a resource system below rates of natural renewal should, one could argue, lead to a sustainable extraction system—but that is a prediction, not a definition. It is, in fact, the foundation of MSY theory (maximum sustainable yield), for many years the basis for management of exploited wildlife and fisheries populations (Roedel 1975). As learned in these fields, a system can only be known to be sustainable after there has been time to observe if the prediction holds true. Usually there is so much uncertainty in estimating natural rates of renewal, and observing and regulating harvest rates, that a simple prediction such as this, as Ludwig *et al.* (1993) correctly observe, is always highly suspect, especially if it is erroneously thought of as a definition.

The second problem is that when one says a system has achieved sustainability, one does not mean an infinite lifespan, but rather a lifespan that is consistent with its time and space scale. Figure 3.2 indicates this relationship by plotting a hypothetical curve of system life expectancy on the *y*-axis versus time and space scale on the *x*-axis.

We expect a cell in an organism to have a relatively short lifespan, the organism to have a longer lifespan, the species to have an even longer lifespan, and the planet to have yet a longer lifespan. But no system (even the universe itself in the extreme case) is expected to have an infinite lifespan. A sustainable system in this context is thus one that attains its full expected lifespan.

Individual humans are sustainable in this context if they achieve their 'normal' maximum lifespan. At the population level, average life expectancy is often used as an indicator of health and well-being of the population, but the population itself is expected to have a much longer lifespan than any individual, and would not be considered to be sustainable if it were to crash prematurely, even if all the individuals in the population were living out their full 'sustainable' lifespans.

Since ecosystems experience succession as a result of changing climatic conditions and internal developmental changes, they have a limited (albeit fairly long) lifespan. The key is differentiating between changes due to normal lifespan limits and changes that cut short the lifespan of the system. Those things that cut short the lifespan of humans are obviously contributors to poor health. Cancer, AIDS, and a host of other ailments do just this.

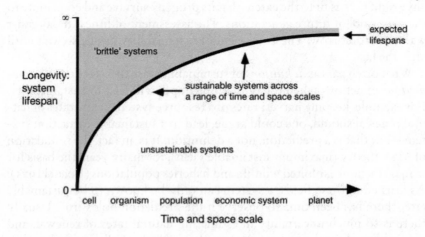

FIG. 3.2. Sustainability as scale (time- and space-) dependent concepts.

Source: Costanza and Patten (1995)

Human-induced eutrophication in aquatic ecosystems causes a radical change in the nature of the system (ending the lifespan of the more oligo-trophic system while beginning the lifespan of a more eutrophic system). We must call this process 'unsustainable' using the above definitions since the lifespan of the first system was cut 'unnaturally' short. The system may have gone eutrophic eventually, but the anthropogenic stress caused this transition to occur 'too soon'.

More formally, this aspect of sustainability can be thought of in terms of the system and the longevity of its component parts.

A system is sustainable if and only if it persists in nominal behavioural states as long as or longer than its expected natural longevity or existence time; and neither component- nor system-level sustainability, as assessed by the longevity criterion, confers sustainability on the other level.

Within this context, one can begin to see the subtle balance between longevity and evolutionary adaptation across a range of scales that is neces-sary for overall sustainability. Evolution cannot occur unless there is limited longevity of the component parts so that new alternatives can be selected. And this longevity must be increasing hierarchically with scale as shown schematically in Figure 3.2. Larger systems can attain longer lifespans because their component parts have shorter lifespans and can adapt to changing conditions. Systems with an improper balance of longevity across scales can become either 'brittle' when their parts last too long and they can-not adapt fast enough (Holling 1986), or 'unsustainable' when their parts do not last long enough and the higher-level system's longevity is cut short.

ECOSYSTEMS AS SUSTAINABLE SYSTEMS

Ecological systems are our best current models of sustainable systems. Better understanding of ecological systems and how they function and maintain themselves can thus yield insights into designing and managing sustainable economic systems. For example, in mature ecosystems all waste and by-products are recycled and used somewhere in the system or fully dissipated. This implies that a characteristic of sustainable economic systems should be a similar 'closing the cycle' by finding productive uses and recycling currently discarded material rather than simply storing it, diluting it, or changing its state, and allowing it to disrupt other existing ecosystems and economic systems that cannot effectively use it.

Ecosystems have had millions of years of trial and error to evolve these closed loops of recycling of organic matter, nutrients, and other materials. A general characteristic of closing the loops and building organized non-polluting natural systems is that the process can take a significant amount of time. The connections, the feedback mechanisms, in the system must evolve, and there are characteristics of systems that enhance and retard evolutionary change. Humans have the special ability to perceive this process and potentially to enhance and accelerate it. The economic system should reinvent the decomposer function of ecological systems.

The earth's first by-product, or pollutant, of the activity of one part of the system that had a disruptive effect on another part of the system was probably oxygen, an unintentional by-product of photosynthesis that was very disruptive to anaerobic respiration. There was so much of this 'pollution' that new species evolved that could use this by-product as a productive input into aerobic respiration. The current biosphere represents a balance between these processes that has evolved over millions of years to ensure that the formerly unintentional by-product is now an absolutely integral component process in the system.

Eutrophication and toxic stress are two current forms of by-products that can be seen as resulting from the inability of the affected systems to evolve fast enough to convert the 'pollution' into useful products and processes. Eutrophication is the introduction of high levels of nutrients into formerly lower nutrient systems. The species of primary producers (and the assemblages of animals that depend on them) that were adapted to the lower nutrient conditions are out-competed by faster-growing species adapted to the higher nutrient conditions. Where the shift in nutrient regime is so sudden that only the primary producers are changed, the result is a disorganized collection of species with much internal disruption (i.e. plankton blooms, fish kills) that can rightly be called pollution. The introduction of high levels of nutrients into a system not adapted to them causes pollution (called eutrophication in this case), whereas the introduction of the same nutrients into a system that *is* adapted to them (i.e. marshes and swamps) could be a positive input. We can minimize the effects of such by-products by finding the places in the ecosystem where they represent a positive input and placing them there. In many cases, what we think of as waste are resources in the wrong place.

Toxic chemicals represent a form of pollution because there are *no* existing natural systems that have ever experienced them and so there are no existing systems to which they represent a positive input. The places where toxic chemicals can most readily find a productive use are probably in other industrial processes, not in natural ecosystems. The solution in this case is to encourage the evolution of industrial processes that can use toxic wastes as productive inputs or to encourage alternative production processes that do not produce the wastes in the first place.

VALUATION OF ECOSYSTEM SERVICES
AND SOCIAL GOALS

Valuation ultimately refers to the contribution of an item to meeting a specific goal or objective. A baseball player is valuable to the extent he contributes to the goal of the team's winning. In ecology a gene is valuable to the extent it contributes to the survival of the individuals possessing it and their progeny. In conventional economics a commodity is valuable to the extent it contributes to the goal of individual welfare as assessed by willingness to pay. The point is that one cannot state a value without stating the goal being served. Conventional economic value is based on the goal of individual utility maximization. But other goals, and thus other values, are possible. For example, if the goal is sustainability, one should assess value based on the contribution to achieving that goal—in addition to value based on the goals of individual utility maximization, social equity, or other goals that may be deemed important. This broadening is particularly important if the goals are potentially in conflict.

There are at least three broad goals that have been identified as important to managing economic systems within the context of the planet's ecological life support system (Daly 1992):

1. assessing and ensuring that the scale or magnitude of human activities within the biosphere are ecologically sustainable;
2. distributing resources and property rights fairly, both within the current generation of humans and between this and future generations, and also between humans and other species; and
3. efficiently allocating resources as constrained and defined by 1 and 2 above, and including both marketed and non-marketed resources, especially ecosystem services.

Several authors have discussed valuation of ecosystem services with respect to goal 3 above—allocative efficiency based on individual utility maximization (e.g. Farber and Costanza 1987; Mitchell and Carson 1989; Costanza et al. 1989; Dixon and Hufschmidt 1990; Pearce 1993; Goulder and Kennedy 1997). We need to explore more fully the implications of extending these concepts to include valuation with respect to the other two goals of (1) ecological sustainability, and (2) distributional fairness (Costanza and Folke 1997). Basing valuation on current individual preferences and utility maximization alone, as is done in conventional analysis, does not necessarily lead to ecological sustainability or social fairness (Bishop 1993).

A Kantian or intrinsic rights approach to valuation (cf Goulder and Kennedy 1997) is one approach to goal 2, but it is important to recognize that the three goals are not 'either–or' alternatives. While they are in some sense independent multiple criteria (Arrow and Raynaud 1986), they must all be satisfied in an integrated fashion to allow human life to continue in a desirable way. Similarly, the valuations that flow from these goals are not 'either–or' alternatives. Rather than a 'utilitarian or intrinsic rights' dichotomy, we must integrate the three goals listed above and their consequent valuations.

A two-tiered approach that combines public discussion and consensus-building on sustainability and equity goals at the community level, with methods for modifying both prices and preferences at the individual level to better reflect these community goals, may be necessary (Rawls 1971; Norton 1995; Norton *et al.* 1998). Estimation of ecosystem values based on sustainability and fairness goals requires treating preferences as endogenous and co-evolving with other ecological, economic, and social variables.

VALUATION WITH SUSTAINABILITY, FAIRNESS, AND EFFICIENCY AS GOALS

We can distinguish at least three types of value, which are relevant to the problem of valuing ecosystem services. These are laid out in Table 3.2, according to their corresponding goal or value basis. Efficiency-based value (E-value) is based on a model of human behaviour sometimes referred to as *Homo economius*—humans act independently, rationally, and in their own self-interest. Value in this context (E-value) is based on current individual preferences that are assumed to be fixed or given (Norton *et al.* 1998). No additional discussion or scientific input is required to form these preferences (since they are assumed to exist already) and value is simply people's revealed willingness to pay for the good or service in question. The best estimate of what people are willing to pay is thought to be what they would actually pay in a well-functioning market. For resources or services for which there is no market (like many ecosystem services) a pseudo market can sometimes be simulated with question-naires that elicit individuals' contingent valuation.

Fairness-based value (F-value) would require that individuals vote their preferences as a member of the community, not as individuals. This model of human behaviour (*Homo communicus*) would engage in much discussion with other members of the community and come to consensus on the values that would be fair to all members of the current and future community (including non-human species), incorporating scientific information about possible future consequences as necessary. One method to implement this

TABLE 3.2 *Valuation of ecosystem services based on the three primary goals of efficiency, fairness, and sustainability*

Goal or value basis	Who votes	Preference basis	Level of discussion required	Level of scientific input required	Specific methods
Efficiency	*Homo economius*	Current individual preferences	Low	Low	Willingness to pay
Fairness	*Homo communicus*	Community preferences	High	Medium	Veil of ignorance
Sustainability	*Homo naturalis*	Whole system preferences	Medium	High	Modelling with precaution

Source: Costanza and Folke (1997).

model might be Rawls's (1971) 'veil of ignorance', where everyone votes as if they were operating with no knowledge of their own individual status in current or future society.

Sustainability-based value (S-value) would require an assessment of the contribution to ecological sustainability of the item in question. The S-value of ecosystem services is connected to their physical, chemical, and biological role in the long-term functioning of the global system. Scientific information about the functioning of the global system is thus critical in assessing S-value, and some discussion and consensus-building is also necessary. If it is accepted that all species, no matter how seemingly uninteresting or lacking in immediate utility, have a role to play in natural ecosystems (Naeem *et al.* 1994; Tilman and Downing 1994; Holling *et al.* 1995), estimates of ecosystem services may be derived from scientific studies of the role of ecosystems and their biota in the overall system, without direct reference to current human preferences. Humans operate as *Homo naturalis* in this context, expressing preferences as if they were representatives of the whole system. Instead of being merely an expression of current individual preferences, S-value becomes a system characteristic related to the item's evolutionary contribution to the survival of the linked ecological economic system. Using this perspective, we may be able to estimate the values contributed by, say, maintenance of water and atmospheric quality to long-term human well-being, including protecting the opportunities of choice for future generations (Golley 1994; Perrings 1994). One way to get

at these values would be to employ systems simulation models that incorporated the major linkages in the system at the appropriate time and space
scales (Costanza *et al.* 1993; Bockstael *et al.* 1995; Voinov *et al.* 1999). To
account for the large uncertainties involved, these models would have to be
used in a precautionary way, looking for the range of possible values and
erring on the side of caution (Costanza and Perrings 1990).

In order to integrate the three goals of ecological sustainability, social
fairness, and economic efficiency we also need a further step, which Sen
(1995) has described as 'value formation through public discussion'. This
can be seen as the essence of real democracy. As Buchanan (1954: 120) put
it: 'The definition of democracy as "government by discussion" implies
that individual values can and do change in the process of decision-making.'
Limiting our valuations and social decision-making to the goal of economic efficiency based on fixed preferences prevents the needed democratic discussion of values and options and leaves us with only the 'illusion of choice'
(Schmookler 1993). So, rather than trying to avoid the difficult questions
raised by the valuation of ecological systems and services, we need to
acknowledge the broad range of goals being served as well as the technical
difficulties involved. We must get on with the process of value formation
and analysis in as participatory and democratic a way as possible, but one
which also takes advantage of the full range and depth of scientific information we have accumulated on ecosystem functioning. This is not simply the
application of the conventional pre-analytic vision and analyses to a new
problem, but will require a new, more comprehensive, more integrated,
pre-analytic vision and new, yet to be developed, analyses that flow from it.

FOUR ALTERNATIVE VISIONS OF THE FUTURE

In addition to changing visions of the way the world works, our vision of
the way we would like the world to be is also changing and evolving.
Elsewhere, I have laid out four broad visions of the future (Costanza 1999,
2000). While there are an infinite number of possible future visions, I
believe these four visions embody the basic patterns within which much of
this variation occurs. Each of the visions is based on some critical assumptions about the way the world works, which may or may not turn out to be
true. This format allows one to identify these assumptions clearly, assess
how critical they are to the relevant vision, and recognize the consequences
of them being wrong.

The four visions derive from two basic world-views whose characteristics are laid out in Table 3.3. These world-views have been described in
many ways (Bossel 1996), but one fundamental distinction has to do with

TABLE 3.3 *Some characteristics of the basic world-views*

Technological optimist	Technological sceptic
Technical progress can deal with any future challenge	Technical progress is limited and ecological carrying capacity must be preserved
Competition is guiding principle	Cooperation is guiding principle
Linear systems: no discontinuities or irreversibilities	Complex, non-linear systems with discontinuities and irreversibilities
Humans dominant over nature	Humans in partnership with nature
Everyone for themselves	Partnership with others
Market as guiding principle	Market as servant of larger goals

Source: Costanza (2000).

one's degree of faith in technological progress (Costanza 1989). The 'technological optimist' world-view assumes that technical progress can solve *all* future problems. It is a vision of continued expansion of humans and their dominion over nature. This is the default vision in our current Western society, one that represents continuation of current trends into the indefinite future. It is the 'taker' culture as described so eloquently by Daniel Quinn in *Ishmael* (Quinn 1992).

There are two versions of this vision, however. One that corresponds to the underlying assumptions on which it is based actually being true in the real world, and one that corresponds to those assumptions being false, as shown in Figure 3.3. The positive version of the technological optimist vision I call Star Trek after the popular US television series which is its most articulate and vividly fleshed-out manifestation. The negative version of the technological optimist vision I call Mad Max after the popular movie of several years ago that embodies many aspects of this vision gone bad.

The 'technological sceptic' vision is one that depends much less on technological change and more on social and community development. It is not in any sense anti-technology, but it does not assume blind faith in technology either. It views technology as the servant of larger goals and seeks to encourage the appropriate *kind* of technology that has the best chance of promoting development without irreversibly damaging our natural capital base. The version of this vision that corresponds to the sceptics being right about the nature of the world I call Ecotopia after the semi-popular book of the late 1970s (Callenbach 1975). If the optimists turn out to be right about the real state of the world, what I call the Big Government vision comes to pass—Ronald Reagan's worst nightmare of overly protective

		Real state of the world	
		Optimists are right (resources are unlimited)	**Sceptics are right** (resources are limited)
World-view and policy	**Technological optimism** Resources are unlimited Technical progress can deal with any challenge Competition promotes progress; markets are the guiding principle	**Star Trek** Fusion energy becomes practical, solving many economic and environmental problems Humans journey to the inner solar system, where population continues to expand (mean rank 2.3)	**Mad Max** Oil production declines and no affordable alternative emerges Financial markets collapse and governments weaken, too broke to maintain order and control over desperate, impoverished populations The world is run by transnational corporations (mean rank –7.7)
	Technological scepticism Resources are limited Progress depends less on technology and more on social and community development Cooperation promotes progress; markets are the servants of larger goals	**Big Government** Governments sanction companies that fail to pursue the public interest Fusion energy is slow to develop due to strict safety standards Family planning programmes stabilize population growth Incomes become more equal (mean rank 0.8)	**Ecotopia** Tax reforms favour ecologically beneficent industries and punish polluters and resource depleters Habitation patterns reduce need for transportation and energy A shift away from consumerism increases quality of life and reduces waste (mean rank 5.1)

FIG. 3.3. Four visions of the future based on the two basic world-views and two alternative real states of the world.

government policies getting in the way of the free market and slowing down economic growth.

Each of these future visions is best described as a narrative from the perspective of, say, the year 2100. This allows one to make them more real and vivid. The narratives are, of course, only caricatures, but they can capture the essence of the visions they represent. I have described these four futures as narratives in detail elsewhere (Costanza 1999, 2000) and here give only a summary of their main features in Figure 3.3.

DEALING WITH UNCERTAINTY AT THE LEVEL OF FUTURE VISIONS

How should society decide among these four visions? Does it even need to decide? Why not just let what happens happen, letting everyone have their own independent vision of the future as it suits them? Is that not the essence of freedom and democracy—everyone being able to pursue their own visions as they please? If everyone lived in their own completely isolated world where their actions and decisions had no effect on anyone else, this might be appropriate. A basic tenet of democracy is that individual rights are not to be limited unless they impact the rights of others. However, we live in a very interconnected world, one which is becoming more and more interconnected every day as the human population grows. All of our futures are intertwined, and the actions and decisions of everyone affect everyone else, both those alive today and those yet to be born. The essence of democracy in this 'full world' context is government by discussion and mutual value formation. The key, as Yankelovich (1991) suggests, is coming to public judgement about the major value issues facing society, its goals and visions, and this process can be accelerated by first laying out the options in the form of relatively well-articulated visions, as I have referred to above.

We can go further in elaborating the consequences of the four visions outlined above in an effort to come more quickly to public judgement. Three of the four visions are 'sustainable' in the sense that they represent continuation of the current society (only Mad Max is not), but one needs to take a closer look at their underlying world-views, their critical assumptions, and the potential costs of those assumptions being wrong.

The world-view and attendant policies of the Star Trek vision imply unbridled faith in technology and free competition, and its essential underlying assumption is unlimited resources, particularly cheap energy. The cost of pursuing this world-view and its policies, if the assumption of unlimited resources is wrong, is something like the Mad Max vision. Likewise, the world-view (and attendant policies) of the Ecotopia vision are technological scepticism and communitarianism (the community comes first), and its essential underlying assumption is that resources are limited and cooperation pays. The cost of pursuing this world-view and its policies if the assumption that resources are limited is wrong is the Big Government vision, where a 'community first' policy slows down growth relative to the free-market Star Trek vision. One can thus think of Figure 3.3 as a 'pay-off matrix'. Each of the four cells in the matrix indicates the 'pay-off' of pursuing the policy and world-view on the left in combination with the real state of the world on the top.

To rank the elements of the pay-off matrix, one would need to discuss the four visions outlined above with a broad range of participants and then have them evaluate each vision in terms of its overall desirability. A preliminary (non-scientific) survey with 418 participants[1] has been

---

[1] The Americans consisted of 17 participants in an ecological economics class at the University of Maryland, 260 attendees at a convocation speech at Wartburg College in Waverly, Iowa, 27 Jan. 1998, and 39 via the World Wide Web. The Swedes consisted of 71 attendees at a Keynotes in Natural Resources Lecture at the Swedish University of Agricultural Science, Uppsala, 20 Apr. 1999, and 31 attendees at a presentation at Stockholm University, 22 Apr. 1999.

conducted. The respondents were read the narrative version of each of the four visions in turn and were then asked: 'For each vision, first state, on a scale of −10 to +10 using the scale provided, how comfortable you would be living in the world described. How desirable do you find such a world? Do not vote for one vision over the others. Consider each vision independently, and just state how desirable (or undesirable) you would find it if you happened to find yourself there.' They were also asked to give their age, gender, and household income range on the survey form. The surveys were conducted with groups from both the United States and Sweden. The results (mean ± standard deviation) are shown in Table 3.4 for each of these groups and pooled.

TABLE 3.4 *Results of a survey of desirability of each of the four visions on a scale of −10 (least desirable) to +10 (most desirable) for self-selected groups of Americans and Swedes*

Future vision	Americans (N=316)	Swedes (N=102)	Pooled (N=418)
Star Trek	+2.38 (± 5.03)	+2.48 (± 5.45)	+2.48 (± 5.13)
Mad Max	−7.78 (± 3.41)	−9.12 (± 2.30)	−8.12 (± 3.23)
Big Government	+0.54 (± 4.44)	+2.32 (± 3.48)	+0.97 (±4.29)
Ecotopia	+5.32 (± 4.10)	+7.33 (± 3.11)	+5.81 (± 3.97)

Note: Standard deviations are given in parentheses after the means
Source: Costanza (2000).

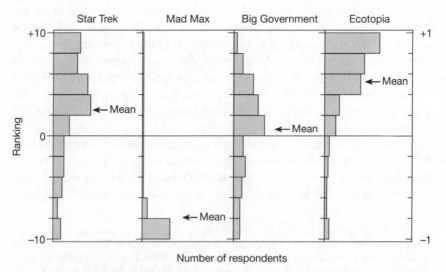

FIG. 3.4. Frequency distributions of responses about the 'degree of desirability' of each of the four future visions. Total number of respondents in each case was 418.

Frequency distributions of the results are plotted in Figure 3.4. The majority of those surveyed found the Star Trek vision positive (mean of +2.48 on a scale from −10 to +10). Given that it represents the logical extension of the currently dominant world-view and culture, it is interesting that this vision was rated so low. I had expected this vision to be rated much higher, and this result may indicate the deep ambivalence many people have about the direction society seems to be headed and/or the self-selected sample of respondents. The frequency plot (and the high standard deviation) also shows this ambivalence towards Star Trek. The responses span the range from +10 to −10, with only a weak preponderance towards the positive side of the scale. This result applied for both the American and Swedish subgroups.

Those surveyed found the Mad Max vision very negative at −8.12 (only about 3 per cent of participants rated this vision positive). This was as expected. The Americans seemed a bit less averse to Mad Max (−7.78) than the Swedes (−9.12), and with a larger standard deviation.

The Big Government vision was rated on average just positive at 0.97. Many found it appealing, but some found it abhorrent (probably because of the limits on individual freedom implied). Here there were significant differences between the Americans and Swedes, with the Swedes (+2.32 ± 3.48) being much more favourably disposed to Big Government and with a smaller standard deviation than the Americans (+0.54 ± 4.44). This also was as expected, given the cultural differences in attitudes towards government in America and Sweden. Swedes rated Big Government almost as highly as Star Trek.

Finally, most of those surveyed found the Ecotopia vision 'very positive' (at 5.81), some wildly so, some only mildly so, but very few (only about 7 per cent of those surveyed) expressed a negative reaction to such a world. Swedes rated Ecotopia significantly higher than Americans, also as might be expected given cultural differences.

Some other interesting patterns emerged from the survey. All of the visions had large standard deviations, but (especially if one looks at the frequency distributions) the Mad Max vision was consistently very negative and the Ecotopia vision was consistently very positive. Age and gender seemed to play a minor, but interesting, role in how individuals rated the visions. Males rated Star Trek higher than females (mean = 3.66 versus 1.90, p (probability of the means being significantly different) = 0.0039). Males also rated Mad Max higher that females (−7.11 versus −8.20, p = 0.0112). The means were not significantly different by gender for either of the other two visions. Age was not significantly correlated with ranking for any of the visions, but the variance in ranking seemed to decrease somewhat with age, with younger participants showing a higher range of ratings than older participants.

Work is in progress to expand this survey, and to conduct a scientific random sample of the entire population, but the general conclusions are fairly insensitive to the exact results.

WORST CASE ANALYSIS

We find ourselves as a species facing the pay-off matrix outlined in Figure 3.3. What do we do? We can choose between the two world-views and their attendant policies. We face pure and irreducible uncertainty concerning the real state of the world. Who knows whether or not 'warm fusion' or something equivalent will be invented? Should we choose the Star Trek vision (and the optimist policies) merely because it is the most popular or because it is the direction in which things seem to be heading already?

From the perspective of game theory, this problem has a fairly definitive answer. This is a 'game against nature' that can only be played once, and the relative probabilities of each outcome are completely unknown. In addition, we can assume that society as a whole should be risk-averse in this situation. The numerical rankings on each outcome (from our preliminary survey) make it a bit easier to talk about:[2] Star Trek is +2.5, Mad Max is −8.1, Big Government is +1.0, and Ecotopia is +5.8. One would look at each row in the matrix (corresponding to a policy set) and see what the worst outcome is for that policy set. For the optimist policy, Mad Max (−8.1) is the worst case. For the sceptical policy set, Big Government (+1.0) is the worst case. One would then choose the policy set with the largest (most positive) worst case. +1.0 is much larger than −8.1, so we would choose the sceptic's policy set. This is a standard 'minimize regrets' decision rule (Milnor 1964; Rawls 1971; Bishop 1978). While there has been some controversy in the literature about the appropriateness of this decision rule in this type of 'game against nature' under extreme uncertainty (Ready and Bishop 1991), a recent review (Palmini 1999) shows that this rule is unambiguously preferred because it 'emphasises risk-aversion while explicitly incorporating the opportunity cost of making a "wrong" choice' (Palmini 1999: 463).

If we choose the sceptic's policy set, the worst thing that can happen is Big Government, which is much better than the worst thing that can happen under the optimist's policy set (Mad Max). The conclusion that we should choose the sceptic's policy set is fairly insensitive to the specific

---

[2] The pooled rankings are used in the discussion, but the conclusions would be the same if using just the American rankings or just the Swedish rankings. In fact, the conclusions are fairly insensitive to the exact values of the rankings as long as Big Government is rated higher than Mad Max, and Star Trek and Ecotopia are rated higher than Big Government.

values of the rankings. The rankings would have to change so that either Big Government or Ecotopia was rated *worse* than Mad Max to reverse this outcome. In fact, the way the pay-off matrix is set up, Mad Max is the one really negative outcome and the one really unsustainable outcome. If one of our major goals as a society is sustainability, then we should develop policies that assure us of not ending up in Mad Max, no matter what happens.

There are also some other considerations in favour of choosing the sceptic's policies. The sceptical policies are less likely to close out any options. One could probably still switch to the optimist's policies, once the real state of the world was shown to conform to the optimist's view. For example, if warm fusion or its equivalent were ever discovered, one could switch to the Star Trek vision from the Big Government vision. The reverse switch from Mad Max to Ecotopia could not be made as easily because the infrastructure would not be there. The sceptic's policies are much better at preserving options.

One could also argue that the probabilities of each state of the world being correct are not completely unknown. If one could argue that the prospects for cheap, unlimited, non-polluting energy were, in fact, very good, then the decision matrix would have to be weighted with those probabilities. But, if anything, the complete dependence of the Star Trek vision on discovering a cheap, unlimited energy source argues for discounting the probability of its occurrence. It is similar to leaping off the top of a tall building and hoping that you can invent a parachute before you hit the ground. Better to wait until you have the parachute (and have tested it extensively) before you jump. By adopting the sceptic's policies, the possibility of this invention is preserved, but without utter dependence on it.

SCIENTIFIC OBJECTIVITY, VALUES, AND POLICY

Ultimately, there is no such thing as 'scientific objectivity' because all science must be based on a pre-analytic vision that is inherently subjective and must be judged for utility and quality against criteria that are inherently subjective. We can, however, be very clear about the distinction between the vision and values component of the process and the analysis component built on that vision.

Scientific work can thus be judged according to its 'quality' based on its adherence to the pre-analytic vision and its pragmatic utility in modelling the real world, as tested against the general criteria developed by the peer community. We can judge between 'good science' and 'bad science' according to these subjectively determined criteria of quality, but it is not honest or useful to use objectivity as a yardstick. Subjective values also enter when we talk about how we would like the world to be. This aspect of future visions strongly determines which set of current policies is most appropriate given our huge level of uncertainty about the real state of the world.

The major source of uncertainty about our current environmental policies is at this level of visions and world-views, not in the details of analysis or implementation within a particular vision. By laying out alternative 'future histories' of the earth, the critical assumptions and uncertainties underlying each vision can be more easily seen, and a rational policy set that assures sustainability can be devised. A cooperative, precautionary policy set that *assumes* limited resources is the most rational and resilient course in the face of fundamental uncertainty about the limits of technology.

REFERENCES

Arrow, K. J., and Raynaud, H. (1986). *Social Choice and Multicriterion Decision-Making*. Cambridge, Mass.: MIT Press.

Berkes, F., and Folke, C. (1994). 'Investing in cultural capital for a sustainable use of natural capital', in A. M. Jansson, M. Hammer, C. Folke, and R. Costanza (eds), *Investing in Natural Capital: The Ecological Economics Approach to Sustainability*. Washington, DC: Island Press.

Bishop, R. C. (1978). 'Endangered species, irreversibility, and uncertainty: the economics of a safe minimum standard'. *American Journal of Agricultural Economics*, 60: 10–18.

Bishop, R. C. (1993). 'Economic efficiency, sustainability, and biodiversity'. *Ambio*, 22: 69–73.

Bockstael, N., Costanza, R., Strand, I., Boynton, W., *et al.* (1995). 'Ecological economic modeling and valuation of ecosystems'. *Ecological Economics*, 14: 143–59.

Bossel, H. (1996). *20/20 Vision: Explorations of Sustainable Futures*. Kassel: Center for Environmental Systems Research, University of Kassel.

Buchanan, J. M. (1954). 'Social choice, democracy, and free markets'. *Journal of Political Economy*, 62: 114–23.

Cairns, J., and Pratt, J. R. (1995). 'The relationship between ecosystem health and delivery of ecosystem services', in D. J. Rapport, C. L. Gaudet, and P. Calow (eds), *Evaluating and Monitoring the Health of Large-Scale Ecosystems*. New York: Springer.

Callenbach, E. (1975). *Ecotopia: The Notebooks and Reports of William Weston*. New York: Bantam.

Costanza, R. (1989). 'What is ecological economics?' *Ecological Economics*, 1: 1–7.

Costanza, R. (1999). 'Four visions of the century ahead: will it be Star Trek, Ecotopia, Big Government, or Mad Max?' *The Futurist*, 33: 23–8.

Costanza, R. (2000). 'Visions of alternative (unpredictable) futures and their use in policy analysis'. *Conservation Ecology*, 4 (1): 5. URL: http://www.consecol.org/vol4/iss1/art5

Costanza, R., and Folke, C. (1997). 'Valuing ecosystem services with efficiency, fairness and sustainability as goals', in Daily (1997).

Costanza, R., and Patten, B. C. (1995). 'Defining and predicting sustainability'. *Ecological Economics*, 15: 193–6.

Costanza, R., and Perrings, C. (1990). 'A flexible assurance bonding system for improved environmental management'. *Ecological Economics*, 2: 57–76.

Costanza, R., Farber, S. C., and Maxwell, J. (1989). 'The valuation and management of wetland ecosystems'. *Ecological Economics*, 1: 335–62.

Costanza, R., Wainger, L., Folke, C., and Mäler, K.-G. (1993). 'Modeling complex ecological economic systems: toward an evolutionary, dynamic understanding of people and nature'. *BioScience*, 43: 545–55.

Costanza, R., d'Arge, R., de Groot, R., Farber, S. *et al.* (1997a). 'The value of the world's ecosystem services and natural capital'. *Nature*, 387: 253–60.

Costanza, R., Cumberland, J. C., Daly, H. E., Goodland, R., and Norgaard, R. (1997b). *An Introduction to Ecological Economics*. Boca Raton: St Lucie Press.

Daily, G. (ed.) (1997). *Nature's Services: Societal Dependence on Natural Ecosystems*. Washington, DC: Island Press.

Daly, H. E. (1992). 'Allocation, distribution, and scale: towards an economics that is efficient, just, and sustainable'. *Ecological Economics*, 6: 185–93.

de Groot, R. S. (1992). *Functions of Nature*. Groningen: Wolters Noordhoff BV.

Dixon, J. A., and Hufschmidt, M. M. (1990). *Economic Valuation Techniques for the Environment: A Case Study Workbook*. Baltimore: Johns Hopkins University Press.

Ehrlich, P. R., and Ehrlich, A. E. (1992). 'The Value of Biodiversity'. *Ambio*, 21: 219–26.

Ehrlich, P. R., and Mooney, H. A. (1983). 'Extinction, substitution and ecosystem services'. *BioScience*, 33: 248–54.

Ekins, P. (1992). 'A four-capital model of wealth creation', in P. Ekins and M. Max-Neef, *Real-Life Economics: Understanding Wealth Creation*. London: Routledge.

Farber, S., and Costanza, R. (1987). 'The economic value of wetlands systems'. *Journal Environmental Management*, 24: 41–51.

Golley, F. B. (1994). 'Rebuilding a humane and ethical decision system for investing in natural capital', in A. M. Jansson, M. Hammer, C. Folke, and R. Costanza (eds), *Investing in Natural Capital: The Ecological Economics Approach to Sustainability*. Washington, DC: Island Press.

Goulder, L. H., and Kennedy, D. (1997). 'Valuing ecosystem services', in G. Daily (ed.), *Ecosystem Services: Their Nature and Value*. Washington, DC: Island Press.

Holling, C. S. (1986). 'Resilience of ecosystem: local surprise and global change', in E. C. Clark and R. E. Munn (eds), *Sustainable Development of the Biosphere*. Cambridge: Cambridge University Press.

Holling, C. S., Schindler, D. W., Walker, B. W., and Roughgarden, J. (1995). 'Biodiversity in the functioning of ecosystems: an ecological synthesis', in Perrings *et al.* (1995).

Knowlton, N. (1992). 'Thresholds and multiple stable states in coral reef community dynamics'. *American Zoologist*, 32: 674–82.

Ludwig, D., Hilborn, R., and Walters, C. (1993). 'Uncertainty, resource exploitation, and conservation: lessons from history'. *Science*, 260: 17–36.

McCoy, R. (1994). *The Best of Deming*. Knoxville, Tenn.: SPC Press.

Milnor, J. (1964). 'Games against nature', in M. Shubik (ed.), *Game Theory and Related Approaches to Social Behavior*. New York: Wiley.

Mitchell, R. C., and Carson, R. T. (1989). *Using Surveys to Value Public Goods: The Contingent Valuation Method*. Washington, DC: Resources for the Future.

Naeem, S., Thompson, L. J., Lawler, S. P., Lawton, J. H., and Woodfin, R. M. (1994). 'Declining biodiversity can alter the performance of ecosystems'. *Nature*, 368: 734–7.

Norton, B. G. (1995). 'Ecological integrity and social values: at what scale?' *Ecosystem Health*, 1: 228–41.

Norton, B., Costanza, R., and Bishop, R. (1998). 'The evolution of preferences: why "sovereign" preferences may not lead to sustainable policies and what to do about it'. *Ecological Economics*, 24: 193–211.

Odum, E. P. (1989). *Ecology and Our Endangered Life-Support Systems*. Sunderland, Mass.: Sinauer Associates.

Palmini, D. (1999). 'Uncertainty, risk aversion, and the game theoretic foundations of the safe minimum standard: a reassessment'. *Ecological Economics*, 29: 463–72.

Pearce, D. (1993). *Economic Values and the Natural World*. London: Earthscan.

Perrings, C. A. (1994). 'Biotic diversity, sustainable development, and natural capital', in A. M. Jansson, M. Hammer, C. Folke, and R. Costanza (eds), *Investing in Natural Capital: The Ecological Economics Approach to Sustainability*. Washington, DC: Island Press.

Perrings, C., and Walker, B. H. (1995). 'Biodiversity loss and the economics of discontinuous change in semi-arid rangelands', in Perrings *et al.* (1995).

Perrings, C. A., Mäler, K.-G., Folke, C., Holling, C. S., and Jansson, B.-O. (eds), (1995). *Biodiversity Loss: Ecological and Economic Issues*. Cambridge: Cambridge University Press.

Quinn, D. (1992). *Ishmael*. New York: Bantam/Turner.

Rawls, J. (1971). *A Theory of Justice*. Oxford: Oxford University Press.

Ready, R. C. and Bishop, R. (1991). 'Endangered species and the safe minimum standard'. *American Journal of Agricultural Economics*, 73: 309–12.

Roedel, P. M. (ed.) (1975). *Optimum Sustainable Yield as a Concept in Fisheries Management*. Special Publication no. 9. Washington, DC: American Fisheries Society.

Scheffer, M., Hosper, S. H., Meyjer, M.-L., Moss B., and Jeppsen. E. (1993). 'Alternative equilibria in shallow lakes'. *Trends in Ecology and Evolution*, 8: 275.

Schmookler, A. B. (1993). *The Illusion of Choice: How the Market Economy Shapes our Destiny*. Albany, NY: State University of New York Press.

Schumpeter, J. (1954). *History of Economic Analysis*. London: Allen & Unwin.

Sen, A. (1995). 'Rationality and social choice'. *American Economic Review*, 85: 1–24.

Senge, P. M. (1990). *The Fifth Discipline: The Art and Practice of the Learning Organization*. New York: Currency-Doubleday.

Tilman, D., and Downing, J. A. (1994). 'Biodiversity and stability in grasslands'. *Nature*, 367: 363–5.

Voinov, A., Costanza, R., Wainger, L., Boumans, R., Villa, F., Maxwell, T., and Voinov, H. (1999). 'The Patuxent landscape model: integrated ecological economic modeling of a watershed'. *Environmental Modelling and Software*, 14: 473–91.

Weisbord, M. (ed.) (1992). *Discovering Common Ground*. San Francisco: Berrett-Koehler.

Weisbord, M., and Janoff, S. (1995). *Future Search: An Action Guide to Finding Common Ground in Organizations and Communities*. San Francisco: Berrett-Koehler.

Yankelovich, D. (1991). *Coming to Public Judgement: Making Democracy Work in a Complex World*. Syracuse: Syracuse University Press.

4

Can Technology Save Us from Global Climate Change?

Bert Metz

THE CLIMATE CHANGE CHALLENGE

BEFORE we can answer the question if technology can save us from global climate change, we need to have an idea how big the climate change challenge is. The most recent IPCC assessment (IPCC 2001a) presents an estimate of the increase in global average mean temperature by 2100 in the absence of further policies to reduce greenhouse gas emissions. On top of the increase of 0.7 degrees Celsius above the pre-industrial global average temperature that has already materialized, a further 1.4 to 5.8 degrees Celsius increase above today's temperature can be expected. This estimate is based on a range of scenarios for the evolution of greenhouse gas emissions developed by IPCC (IPCC 2000a), combined with a range of climate models and a range of climate sensitivities.[1] The emissions scenarios are obviously where we have to look when we want to get an idea of the amount of greenhouse gas emissions to be avoided.

The latest IPCC scenario set, described in the *Special Report on Emissions Scenarios* (*SRES*, see IPCC 2000a), covers a wide range of possible futures. It explores various future worlds with differences in value systems (economic versus social–environmental emphasis) and market orientation (globalization versus regional focus) against the background of various population projections, economic growth assumptions, and technology and land use choices. The possible future societies act as a context to ensure a coherent set of assumptions on these critical variables.

---

[1] Climate sensitivity is the degree of warming for a doubling of CO_2 concentrations in the atmosphere; values vary between 1.5 and 4.5 °C and reflect uncertainties about the feedbacks in the climate system due to cloud formation, water, etc.

These assumptions were used as inputs by six integrated assessment models to calculate projected emissions. The result is a range of emissions of
greenhouse gases in 2100 from about current global levels to a fourfold
increase.

The next question is 'Below what level do the emissions trajectories have
to be kept in order that climate change is still tolerable?' This is a political
decision. Article 2 of the United Nations Framework Convention on
Climate Change (*UNFCCC* 1992), which is the basis for international
action to combat climate change, specifies as its objective to stabilize
greenhouse gas concentrations in the atmosphere at such a level and at such
a rate that dangerous anthropogenic interference with the climate system is
avoided. It emphasizes nature protection, food security, and the possibility
of sustainable development as the most critical areas of possible dangerous
interference. This means, for instance, that a stabilization level that seriously affects biodiversity or leads to major shifts in food production
should be avoided. Also rapid changes of the climate over a certain period
could lead to comparable problems even at lower concentration levels.
With respect to sustainable development there are two sides to the coin:
significant climate change and the impacts of it can be economically and
socially very harmful and are therefore not compatible with sustainable
development. However, rapid changes in economies to drastically reduce
greenhouse gas emissions can be economically and socially disruptive as
well and that could also negatively affect sustainable development.

'Danger' is a subjective notion. Science can provide information on the
impacts of climate change belonging to various stabilization levels. It can
also provide information on what is needed to achieve those levels and
what such action would cost. It can also show the costs of slower or more
rapid mitigation. But the decision on what is dangerous is not a scientific
but a political one.

The latest IPCC assessment report (IPCC 2001*b*) presents the best
available information on the impacts of climate change. It shows that
impacts of the already changed climate are visible today and that risks will
increase significantly in the future. It provides information that makes it
possible to link stabilization levels with global average temperature
increase and climate change impacts. Summarizing this information
including regional effects and sensitive ecosystems makes clear that at
more than around 1–2°C above current temperatures these risks are
becoming significant.

On the question of cost of mitigation (the other element of the determination what is dangerous) the IPCC *Third Assessment Report* shows
that total costs (over a fifty- to hundred-year period) of stabilization
increase significantly with lower stabilization levels. However, even for

the lower stabilization levels, annual costs in terms of lower GDP growth are generally very small, less than 0.1 per cent, provided the time is used to make the changes at the economically optimum moment (Hourcade *et al.* 2001). It also shows that for low-level stabilization, given the need for global emissions to start declining within the next few decades (Prentice *et al.* 2001), the inertia in the economic system, and assuming low discount rates, postponing mitigation action could very well increase costs. Studies that investigate the effect of uncertainty in the eventual stabilization level also show an advantage of anticipating a possible low-level stabilization decision by making early reductions (Tóth *et al.* 2001).

So, let us now make an estimate of the climate change challenge, i.e. the amount of greenhouse gases to be avoided. Starting from a selected 'tolerable' risk level, the corresponding temperature and greenhouse gas concentration stabilization level is found, using climate models and assuming a value for the climate sensitivity. Knowing the stabilization level, an emissions profile leading to this stabilization level can be calculated with carbon cycle models. For all stabilization levels these emission profiles show eventually a sharp decline to far below current values. The lower the stabilization level, the earlier this has to occur. This is caused by the long lifetimes of many greenhouse gases in the atmosphere. We can compare these stabilization profiles with emission profiles for a situation where no action is taken, the so-called reference scenarios as, for instance, covered by the IPCC *SRES* (IPCC 2000a). Depending on the chosen stabilization level and the assumed future (reference scenario) a 'gap' in emissions can be calculated. Because of the long lifetimes of many greenhouse gases the cumulative emissions over the hundred-year time-frame can be used to express the magnitude of this 'gap'.

For the purposes of this exercise we assume that the stabilization level chosen would be 450 parts per million, by volume (ppmv) CO_2. This is not an unreasonable assumption because for a 450 ppmv CO_2 stabilization global average temperatures in 2100 will be 1.2–2.3°C above today[2] and in the long term, after the climate system has reached equilibrium again, 1.4–4°C above today (IPCC 2002). At that level, impacts can be serious, and mitigation costs do not seem to be excessive if well planned. The cumulative amount of carbon equivalent emissions to be avoided over the coming hundred years would be about 300–1,500 GtCeq.[3] The lower end of this range is for societies that in the future already move towards sustainable

---

[2] A stabilization at 450 ppmv CO_2, together with the contribution from other greenhouse gases, would lead to a further temperature increase of about 1.5°C above current values given a medium value for the climate sensitivity.

[3] Carbon equivalent emissions are the sum of all greenhouse gas emissions, weighted according to their heat-trapping contribution (global warming potential).

development; the higher end of the range is for societies with high fossil fuel use (Morita *et al.* 2000, 2001).

In the subsequent paragraphs this range of 'emissions to be avoided' will be the reference when investigating the role that technology can play to realize these reduced emissions.

THE TECHNOLOGICAL POTENTIAL FOR AVOIDING GREENHOUSE GAS EMISSIONS

When discussing the potential of technologies to realize emission reductions, it is important to make a distinction between what is technically feasible and what the market accepts. The following potentials can be distinguished (Sathaye *et al.* 2001):

- the *technological potential*: what has technologically been demonstrated at practical level;
- the *socio-economic potential*: the part of the technological potential that is realizable given social preferences and public rate of return requirements;
- the *economic potential*: the part of the socio-economic potential that is economically attractive given current prices and private sector rate of return requirements;
- the *market potential*: the part of the economic potential that is actually realized under current market conditions.

Changes in prices, policy interventions, and other measures can bring the market potential closer to the technological one. In the next section I will investigate the barriers and ways to overcome them, but first an estimate of the various potentials will be given.

Short- to Medium-Term Options

In the short to medium term (ten to twenty years from now) there is a very significant potential for emission reduction. The IPCC *Third Assessment Report* gives the most recent overview (Moomaw *et al.* 2001). When all technological options that can reduce greenhouse gas emissions are taken whose net direct costs are below \$100/tCeq[4] then global emissions in the

---

[4] Net direct costs are direct costs (capital, operating, and maintenance costs) minus direct benefits (energy saved). Additional implementation costs and ancillary benefits for other environmental problems or the economy are not included. Costs are expressed in 1998 US dollars per tonne of reduced gases on a carbon-equivalent basis, using rate of return requirements of 5–12% per year or payback periods of eight to twenty years.

2010–20 time-frame can be reduced to below current levels, while trends would otherwise lead to sharp increases. This potential is closest to the socio-economic potential identified above. Half of this potential is even available at net benefits, i.e. the energy costs saved are higher than the costs of taking the measure. Current market conditions would not lead to such reductions. Policy interventions would therefore be needed.

The technologies involved in this short- to medium-term potential (at costs below \$100/tCeq) include: technologies that improve energy efficiency; renewable energy technologies (wind energy, bio-energy); technologies to capture and store CO_2, new vehicle engine designs (fuel cells); new process designs in industry eliminating release of greenhouse gases like nitrous oxides or perfluorocarbons; enhanced fixation of CO_2 in forests and soils; waste recycling and methane recovery from waste disposal sites; and replacing products like hydrofluorocarbons (alternatives for banned CFCs). Energy efficiency improvement is by far the most important in this time-frame given its relatively low costs.

The building sector (residential housing and commercial buildings, which account for about 30 per cent of current greenhouse gas emissions) in particular stands out as promising for strong energy efficiency improvements: more than half of current emissions in this sector can be eliminated by 2020 at negative direct costs. In the industry sector (manufacturing of raw materials and goods, representing almost 45 per cent of current emissions) a combination of energy efficiency, using less materials and options to eliminate or reduce emissions of non-CO_2 greenhouse gases, can also lead to emission reductions. Reductions can go up to about 60 per cent of current emissions, but at costs rising to \$100/tCeq. The transport sector shows somewhat less potential in the medium term. By 2020 an emission reduction of 40 per cent seems feasible below a cost of \$100/tCeq, but at the same time this is the fastest-growing sector and it is a sector in which low-cost opportunities are often not used. Over the period 1990–5 emissions grew at 2.4 per cent per year and that steep growth rate does not show signs of going down.

The total emission reductions are estimated at 1.9–2.6 GtCeq per year by 2010 and 3.6–5.0 GtCeq per year by 2020. Compared to the amount of 300–1,500 GtCeq to be avoided over the period 2000–100 this is still insufficient. However, much stronger reductions are possible over a longer time-frame than the next ten to twenty years.

Long-Term Options

The long-term potential (fifty to a hundred years) is much harder to express in economic terms. Cost figures for technologies over such a long time period are difficult to estimate. Therefore, we have to rely on estimates for the technological potential. Based on currently available technologies it is possible to demonstrate a huge technological potential.

For instance, there are several technical alternatives available for gasoline or diesel-driven combustion engine cars, such as changing to compressed natural gas as fuel, adding biofuel to gasoline,[5] hybrid gasoline–electric vehicles, fuel-cell-driven vehicles, or all-electric vehicles. Except for fuel cell vehicles, which are expected to enter the market by 2005, the other alternatives are in practical use today at a small scale. Emissions of CO_2 per kilometre travelled vary enormously across these alternatives. For electric traction the fuel mix from which the electricity to run the vehicle is produced makes a difference. Hydro-power or other renewable energy supply options can effectively reduce emissions from electric vehicles to very low levels, while all-coal electricity generation would hardly lead to an advantage over gasoline-fuelled vehicles. Fuel cells produce electricity directly by combining hydrogen and oxygen; the hydrogen can be produced on board the vehicle from gasoline or methanol, in which case the carbon emissions are only reduced moderately or can be produced outside the vehicle in decentralized (gas stations) or centralized facilities. The latter option opens the possibility of producing hydrogen from fossil fuel (currently the standard method in the chemical industry to produce hydrogen) and eliminating CO_2 release by capturing CO_2 at the point of origin and locking that away in empty oil or gas fields or deep aquifers. In the long term hydrogen could also be produced through solar energy. Comparing the various alternatives shows that there is the technological potential with currently available technologies to reduce vehicle emissions up to 90–5 per cent (Moomaw *et al.* 2001).

Electricity for residential and industrial use can be supplied by renewable sources: hydro-power, geothermal, wind, ocean, solar, and biomass. There is a big technological potential for these renewable energy sources, even if we take account of the limitations of available land and the intermittent character of some of these sources (in particular wind and solar). The total supply is estimated to be about 4,200 EJ.[6] This is much bigger than the top end of the range of total world primary energy demand according to the range of IPCC scenarios by the year 2100 (500–2,750EJ) (IPCC 2000*a*). Solar and biomass energy are by far the largest in this portfolio (2,600 and 1,300 EJ respectively). In other words, in the electricity

---

[5] Biofuel such as ethanol from agriculture residues or grains or diesel fuel from oilseeds has an approximately zero net emission because CO_2 released from combustion is taken up again by the agricultural crops where it originates from.

[6] Energy use is expressed in terms of Exajoule (EJ), equivalent to 10^{18} J.

sector the technical potential is available to make huge reductions in greenhouse gas emissions. When we also take into account nuclear energy, this could add another 100–4,000 EJ to the mix of low-CO_2 energy supply options[7] (IPCC 2000*a*).

The capture and storage of CO_2, already touched upon above in the discussion of fuel cell vehicles, is an important technological option to reduce emissions. The technology is well known: hydrogen production today in the chemical industry from fossil fuel involves a chemical process in which coal, oil, or gas is turned into hydrogen and CO_2, and the CO_2 is separated from the hydrogen. Currently most of this CO_2 is vented to the atmosphere, but some has been captured for various industrial uses. As far as storage of CO_2 is concerned, pumping into oil or gas fields is a prime option that has already been used for a long time to increase the pressure in these oil or gas fields. There is at least one place where powerplant-related CO_2 is being used now (Wilson *et al.* 2001). Other storage options are deep aquifers, unmineable coal beds, and deep ocean disposal (Williams *et al.* 2000). There is currently a natural-gas-processing plant in operation in Norway, where CO_2 removal and storage in a deep aquifer is applied commercially (Baklid and Korbol 1996). Admittedly, the reason for this commercial viability is the combination of a high CO_2 natural gas field and the presence of a tax in Norway on CO_2 emissions that made it attractive to store the CO_2 that had to be separated underground rather than venting it to the atmosphere. Estimates of the total storage capacity available indicate that about 1,500 GtC could be stored in empty oil and gas fields and deep aquifers, an encouraging conclusion given the 300–1,500 GtC that needs to be avoided over the coming hundred years. Unmineable coal bed absorption, a technology that now has successfully been demonstrated (Stevens *et al.* 1999), and deep-sea disposal, whose potential environmental impacts are not yet well known, could add another 1,000 GtC or more (Herzog *et al.* 2001). The security of the stored CO_2, in particular in more 'open destinations' such as deep oceans or aquifers, is an important issue that needs further research. Current cost of CO_2 removal and storage away from the atmosphere varies between \$150–220/tC, but there are many technical opportunities that may reduce costs in the future (Moomaw *et al.* 2001; Williams *et al.* 2000).

Another possibility is the biological sequestration of CO_2 by increased uptake in vegetation (forests in particular) and soils or reduced release of carbon from deforestation and other land use changes (conservation). The total amount of carbon emissions that can be avoided or compensated in

[7] The nuclear energy potential varies with assumptions about nuclear fuel reprocessing. Other aspects of nuclear energy such as the risk of reactor incidents, nuclear waste processing and storage, and proliferation of nuclear arms materials are not considered here.

this way over the next fifty years or so is estimated to be about 100 GtC. Substitution of fossil fuel by using biomass is often more attractive than just keeping the carbon locked into wood, but is already counted as part of the biomass potential indicated above. The largest potential for this mitigation option is in subtropical and tropical regions, where land and water availability and the adoption of different land management practices might be problematic in certain areas. Current cost estimates for these biological mitigation options have a high degree of uncertainty. The literature reports figures ranging from $0.1/tC to $20/tC for tropical countries and $20/tC to $100/tC for non-tropical countries. Differences among calculation methods, omission of costs for infrastructure and monitoring, and omission of opportunity cost of land are among the major factors explaining these differences. Many of the biological options may have other social, economic, and environmental benefits that generally have not been taken into account. There are also risks of worsening existing problems like loss of biodiversity, groundwater extraction, or pollution and loss of community cohesion. Land use change practices for carbon management can have a significant effect on emissions of other greenhouse gases. Depending on the situation, this could lead to either increased or decreased emissions of other greenhouse gases (IPCC 2000*b*; Kauppi *et al.* 2001).

Non-CO_2 greenhouse gases emissions, mainly methane, nitrous oxides, and halocarbons, are modest compared to CO_2, but due to their higher warming effect per unit of mass they have contributed about 40 per cent to increased radiative forcing of greenhouse gases over the period 1750 to present; and it is radiative forcing that drives climate change. By 2100 the dominance of CO_2 will have increased according to the IPCC *SRES* scenario range and the contribution of non-CO_2 gases would be in the order of 20–30 per cent (Ramaswamy *et al.* 2001). The manufacturing industry, waste management, agriculture and land-use change are the main sources. Technically, virtually all non-CO_2 emissions from industry and waste management can be eliminated, and in other sectors strong reductions are possible. No reliable estimates exist, however, to give more precise figures (Moomaw *et al.* 2001). This means that reduction of these gases could further contribute to the challenge of reducing the 300–1,500 GtCeq cumulative emissions, making a contribution of something in the order of a few hundred GtCeq to the avoidance of emissions.

HOW DOES IT FIT TOGETHER?

An adequate technological potential to reduce greenhouse gas emissions is not sufficient. The cost of these options has to come down significantly in

order to make them a serious competitor to traditional practices. What are the perspectives for cost reductions in the long term? There is ample evidence from past experiences that the cost of new technologies can come down strongly given enough time (Gruebler 1998; International Energy Agency 2000). There appears to be a strong relationship between costs and (cumulative) numbers of installations or capacity of technologies installed. The experience gained with implementing commercial installations is extremely valuable for further improvements and cost reductions. The revenues from sales of earlier versions of technologies help to finance development of improved versions. A clear example of this phenomenon is the cost of electricity production from wind turbines in Europe. Over the period 1980–95 the price of electricity from wind came down by about a factor of 10 to a level that is now competitive with natural gas combined cycle power generation and advanced coal combustion plants that are the standard fossil-fuel-based installations of choice. The same phenomenon is happening for solar and biomass electricity, but the absolute cost levels are still much higher than that of standard fossil technology (a factor of 10 and 3 to 4 respectively). The rate at which the costs of solar electricity come down is fast, however: for each doubling of installed capacity the price comes down by a factor of about 3. For wind power this is now a factor of about 1.2, and for natural gas combined cycle plants a factor of slightly more than 1. This means that renewable energy technologies most likely will become competitive with traditional fossil-fuel-based technologies in the medium to longer term. This mechanism of 'technological learning' also applies to other greenhouse-gas-reducing technologies. It supports the expectation that the economic potential of these low-carbon technologies can become significant in the long term.

What will happen with fossil-fuel-based technologies over the long term? Until recently the common wisdom was that fossil fuel would run out and that this would naturally lead to strong increases in prices that could further pave the way for new technologies. The current assessment of fossil fuel resources in comparison to energy demand projections, however, shows abundant resources.[8] Even without counting huge quantities

---

[8] Reserves are those occurrences that are identified and measured as economically and technically recoverable with current technologies and prices. Resources are those occurrences with less certain geological and/or economic characteristics, but which are considered potentially recoverable with foreseeable technological and economic developments. The resource base includes both categories. On top of that, there are additional quantities with unknown certainty of occurrence and/or with unknown or no economic significance in the foreseeable future, referred to as 'additional occurrences'. Examples of unconventional fossil fuel resources include tar sands, shale oil, other heavy oil, coal bed methane, deep geopressured gas, and gas in acquifers.

of methane clathrates[9] the total coal, oil, and gas resources, both conventional as well as unconventional ones, amount to about 5,000 GtC (Moomaw *et al.* 2001). For the most fossil-fuel-intensive IPCC scenario an amount of 2,100 GtC would be burned over the next hundred years, so availability of resources does not seem to be a major constraint. The environmental and climate consequences may in fact be a much more serious limitation for fossil fuel use. For a stabilization level of 1,000 ppmv CO_2 in the atmosphere (at which level global average mean temperature could increase between 3.5°C and 9°C above today's level) the amount to be released over the coming hundred years would be about 1,500 GtC (IPCC 2001c). There is enough fossil fuel to make that possible.

Would the price of fossil fuel be a natural limitation? It is likely that the price of fossil fuel will go up, particularly for oil and gas when the most accessible sources have been exhausted. Given enough investment, unconventional fossil fuel resources are likely to be available at much lower costs than today, meaning that fossil fuel costs would not necessarily rise sharply. This would not, however, exclude temporary shortages due to political circumstances.

In addition to the upward trend in fossil fuel prices and the beneficial effects of technological learning (making traditional fossil-based technologies relatively less attractive than renewable energy technologies) the relative attractiveness of new technologies will also be affected by policy interventions. There is a large range of policies and measures that can be introduced to make the application of new technologies more attractive (Bashmakov *et al.* 2001). Basically, these policies and measures make it attractive to invest in emission reductions (because it is legally required or commercially attractive), and make traditional fossil fuel technologies relatively more expensive and low-greenhouse gas technologies relatively less costly.

Modelling studies show that realistic assumptions about the various components that drive long-term change, such as the rate at which new technologies can be introduced, the technological learning effect, and the increase in price of fossil fuels, can indeed lead to drastic greenhouse gas emission reductions over the next fifty to a hundred years. It would allow to stabilize atmospheric CO_2 concentrations at 450 ppmv by 2100 against both an A1 as well as a B1 reference scenario (Morita *et al.* 2000, 2001; Schellnhuber and Held, Ch. 1 in this volume; van Vuuren and de Vries 2001). This is equivalent to a cumulative emission avoidance of 300–1,500

---

[9] Methane clathrates (natural gas hydrates) occur in quantities of about 12,000 GtC, which is more than twice the amount of all gas, oil, and coal reserves and resources combined. The costs at which these clathrates are available are uncertain (Moomaw *et al.* 2001).

GtC. For an A2 reference scenario, however, the circumstances (including a fragmented world, high reliance on coal, low economic growth, low innovation potential, low environmental awareness) seem prohibitive for reaching these low carbon emission levels.

In terms of required investments, the 450 ppmv stabilization case referred to above (against the background of a B1 reference scenario) does require additional investments, especially in the energy sector. Investments will not, however, go beyond the levels over the past thirty years, when they averaged 2–3 per cent of GDP. This is because there is also a shift of energy investments, from traditional energy sources into renewable energy technologies (van Vuuren and de Vries 2001).

As indicated above, in order to reach a 450 ppmv CO_2 stabilization level global emissions need to peak within a few decades from now and to go below current levels by around 2035 or so (Prentice *et al.* 2001). This implies that investments have to be avoided that would further lock the economic system into a high emissions mode[10] and that cost-effective options to reduce greenhouse gas emissions need to be implemented. It also requires that innovation is strongly promoted to pave the way for the penetration of new technologies, new technology infrastructures (such as hydrogen pipelines to supply automobile and other fuel cells), and 'paradigm shifts' (such as telecommuting and internet-based business operations). These 'transition strategies' have to be managed carefully in a participatory manner, i.e. government, business community, and social stakeholders need to be involved in order to get maximum synergy (Rotmans *et al.* 2001).

As far as macro-economic costs of such reductions are concerned, the GDP reduction at the end of a fifty- to a hundred-year period of emission reduction policies is estimated to be a few percentage points below what it would have been without climate policy. This is assuming the most efficient way of implementation is chosen, i.e. the lowest cost options are implemented first and premature capital retirement is avoided as much as possible. For the range of IPCC *SRES* scenarios in which absolute GDP levels increase strongly over a fifty- to a hundred-year period, GDP reductions due to stabilization policies do not lead to significant declines in GDP growth rates over that period. The annual 1990–2100 GDP growth rate, on average across all scenarios, was reduced by only 0.003 per cent per year, with a maximum reaching 0.06 per cent per year. Mitigation costs may, however, be higher for some regions and during some periods (Hourcade *et al.* 2001). So, except for a very 'climate-change-unfriendly' world, it

---

[10] The US Energy Plan as proposed by President George W. Bush can be characterized as such in view of its emphasis on investments in domestic high-carbon-intensive fossil fuel resources.

does seem realistic to assume that the technological potential can be realized to address the climate change challenge.

THE ROLE OF POPULATION, ECONOMIC GROWTH, AND HUMAN BEHAVIOUR

Would the effect of technological improvement not be completely overshadowed by increasing population, economic growth, and choices for an energy- and material-goods-intensive lifestyle? It is a well-known phenomenon today that improvement in environmental performance of, for instance, motor vehicles is more than compensated by a growing population, more vehicle ownership, more vehicle use as incomes grow, and shifting preferences towards bigger and less fuel-efficient cars. The estimated amounts of emissions to be avoided described above, however, have been calculated taking into account population growth, economic growth, and certain social and human preferences. They are based on the IPCC *SRES* scenarios that explicitly include these variables. So, using these emission scenarios ensures that all the future developments that counterbalance the use of more efficient technologies are included. The upper end of the 300–1,500 GtCeq range of 'emissions to be avoided' reflects these high emission scenarios. The lower end of the range reflects the low emission scenarios that assume stronger preferences for social and environmental values and a lower economic growth rate (population assumptions happen to be the same for both the high and the low end of the scenario range). In other words, lifestyle choices and economic growth do have a very significant impact on the amount of emissions that need to be avoided. It would therefore help a lot if society moved in a more sustainable direction in the future. But it is not strictly needed, because even with a strong material wealth orientation and high economic growth (the IPCC *SRES* A1 world) the potential is available to achieve the necessary emission reductions.

WHAT DOES IT TAKE TO REALIZE THE NECESSARY EMISSIONS REDUCTIONS?

The sections above have shown that all the rational arguments about technological possibilities, rates of improvement, changing prices over time, economic costs, and availability of policies and measures point towards the climate change challenge being surmountable. The ultimate question is 'can this technological potential be realized in practice?' Implementing policies, changing decision-making, and changing human behaviour is far

from easy. And the changes will have to happen at the right time and relatively soon. What are the barriers that will have to be overcome? What
policies and measures can help overcome these barriers?

When looking at the challenges of realizing the timely introduction of
new technologies, it is important not only to think about industrialized-
country situations. Much of the future emissions that need to be avoided
are in developing countries. That is where the biggest growth in emissions
will occur, given the increase in welfare necessary to eliminate poverty and
improve living conditions. Depending on the baseline scenario, the developing-country share of avoided emissions would be in the order of 60–80
per cent. So that is also where new technologies have to be introduced on a
large scale. This brings in a whole range of potential barriers related to
social change in developing countries and transfer of technology.

Barriers

Barriers can be analysed effectively in the context of the different categories
of potentials introduced above. Barriers are defined as the (many) problems
in moving the potential up towards the technological potential (Sathaye *et al.*
2001). Using the categorization of the IPCC Special Report on Technology
Transfer (IPCC 2000*b*), five classes of barriers can be distinguished:

Human capacity constraints. These are most prominent in developing
countries, where there is an enormous need for education and training.

Organizational capacity constraints. Again, these are most prominent in
developing countries, where there is a rather poor availability of all sorts of
supporting organizations, such as legal and marketing services, management consultants, industry organizations, etc. The absence of a tradition of
involving private sector interest groups, citizen organizations, and environmental organizations often leads to projects that are not seen as appropriate by stakeholders. A problem that also occurs in industrialized
countries is lack of supply channels for new technologies and lack of
appropriate technical services for installation and maintenance of these
new technologies.

Information constraints. In many countries, but especially in developing countries, relevant information about new technologies is not (or is
not easily) available. Although information systems have been set up in
many places and even internationally (nowadays mostly with internet
access), many of these information systems are ineffective because they
are not updated sufficiently, are difficult to access, or are not sufficiently
focused on the needs of actual users. Another important factor is that
information about new technologies cannot easily be evaluated owing to
lack of technical standards that include these new technologies. Where

information about technologies and their cost-effectiveness is available, it is often not used because energy and greenhouse gas emission savings are not receiving adequate attention by management. In developing countries the general lack of research institutes to help assess new technologies forms an additional constraint.

Enabling environment. There are a number of factors that determine whether or not it is attractive for private (and public) investors to produce or market new technologies. Many are economic: the price of fossil-fuel-based energy is often subsidized, and environmental and climate damage of emissions from fossil fuel use are not factored into the price. In developing countries the investment risk might be seen as too high owing to macro-economic instability, lack of patent protection, limitations to transfer of profits, competition with imported technologies that are subsidized by the exporting countries, etc. Non-economic factors are, for instance, the resistance by vested interest, corruption, import restrictions, and problems to obtain licences.

Social and behavioural obstacles. Particularly in industrialized countries, social and behavioural constraints are important. Even if cheaper and more environmentally friendly possibilities exist, people often do not choose them because of personal preferences. With rising incomes, cost-effectiveness of transportation or more efficient energy use becomes less of a concern. Aesthetic considerations can overrule cost-effective options (Sathaye *et al.* 2001). An example of the latter is the widespread resistance of large numbers of people against wind turbines in the Netherlands (a country with a long history of windmills that are still a trademark for the country) because they spoil the flat open landscape. Another example: consumers in industrialized countries avoid buying compact fluorescent lamps for aesthetic reasons, although they are economically very attractive.

Different combinations of barriers are responsible for limitations to the market potential of technologies. They vary strongly between countries and sectors of the economy.

Policy Interventions

Before focusing on specific climate change and technology related policy options it is important to stress that socio-economic, trade, and other non-climate policies could have a strong influence on creating the conditions for innovation, technology transfer, and implementation. Policies to eliminate or reduce macro-economic instability, the influence of vested interests, corruption, import restrictions, limitations to transfer of profits, subsidies by exporting countries, lack of patent protection, and problems to obtain licences are examples of this. The priority policies and measures directly focusing on increasing the market potential for technologies to address climate change vary from region to region. A distinction is made between OECD countries, countries in central and eastern Europe that are going through a process of transition to a market economy (countries with economies in transition), and developing countries.

Industrialized (OECD) Countries

Full-cost pricing: Since many investment decisions in these countries are market-driven, it is vital to have prices reflect the real costs, i.e. including environmental and social damages (externalities). Only then can environmentally or socially superior options get a real chance to penetrate. Of course, this includes abolishing subsidies that further distort the real prices, such as on fossil fuel or fossil-fuel-based electricity (van Beers and de Moor 2001). In many cases government-imposed taxes are a means of approximating such a full cost price.

Research and development: New technological developments very often originate in OECD countries that have the intellectual, financial, and organizational basis. Public R & D funding for energy technologies, however, has declined seriously over the last twenty years, in absolute and per unit of GDP terms (International Energy Agency 2001). Private R & D in this sector has suffered from the energy market liberalization trends. Increasing R & D funding and cooperative R & D programmes with institutions in developing countries would be important.

Removing social and behavioural barriers: Human behaviour is more than just responding to cost minimization incentives. Convenience, time, social status, beliefs, and habits are often more important. This explains, for instance, why economic energy efficiency potentials in households or transportation are difficult to achieve. If policy interventions can be designed to maximize additional benefits in terms of health, family values, or social status, they may be much more effective. Raising awareness of the consequences of climate change and of the possibilities that exist to address those is an important factor. Better understanding the motivations of people and the social pressures to which they are sensitive would also help. New values and identities emerge such as 'green consumption' and those can be built upon. Broad participation of people in decision-making processes could be a way to mobilize social innovation (Sathaye *et al.* 2001; Jochem *et al.* 2001). Knowing the complexities of human choice and behaviour, use of regulatory rather than market instruments may be appropriate in cases where this could be a major obstacle.

Political determination: Overcoming pressure from lobby groups that resist changes is probably one of the most difficult things for political decision-makers. The recent 'fuel price revolt' in Europe (Mitchell and Dolun 2001), where massive social protest against rising gasoline and diesel prices led many European governments to reduce fuel taxes or provide other forms of financial compensation to the transport sector, is a good example of succumbing to such pressures. Political leadership is needed to capture the opportunities for economically or socially justified change. It means ample attention should be given to assist those that tend to lose from changes to adjust to the new circumstances. Especially, social and employment concerns need to be taken very seriously.

Countries with Economies in Transition

Full-cost pricing: The importance of full-cost pricing is even greater in countries with economies in transition. Restructuring of the economy and of the governance structure has not yet been fully completed in many of these countries. Declining income levels are reasons for maintaining subsidies. Imposing energy taxes is politically unattractive, even apart from the problems of collecting the tax. Many investment decisions need to be taken that will determine energy and transportation infrastructure for a long time to come. Environmental friendly technologies then have to be given a fair chance.

Awareness: Environment and climate change are not of great concern for people in countries with economies in transition at present. Before the big changes in the political structure in central and eastern Europe environmental movements were strong. However, they channelled a lot of social protests. After the change, much of that social force shifted to social and economic concerns, making environment and climate change a politically unimportant issue. Building awareness about climate change and the opportunities to combat it is therefore of crucial importance for this group of countries.

Developing Countries

Full-cost pricing: The main contribution to projected greenhouse gas emissions from developing countries is typically the result of activities in the future and installations that still have to be built. What was said on the inclusion of environmental and climate change damages in the price of fossil-fuel-based energy for countries with economies in transition therefore applies even more strongly for developing countries. However, there are significant pressures in developing countries to keep energy prices low and to provide subsidies, even to environmentally harmful practices such as unsustainable agriculture and forest exploitation (van Beers and de Moor 2001).

Access to information: Developing countries have much to gain from better information about new technologies. It is not just a matter of creating yet another computerized international database with subsidized access for organizations in developing countries, although expanding existing systems is certainly useful (CADDET 1998). Much depends on who is making choices on technology. In the case of foreign direct investment, via either subsidiaries or joint ventures, choices are mostly made by the investors that could include local private or public entities. Ensuring adequate local R&D capacity—in the private and public sector—to evaluate proposed technology choices and suggest adaptation of technologies to local circumstances can be an effective step to help make the right choices. Professional associations, specialized companies, conferences, and other formal or informal networking activities generally play an important role in spreading information (McKenzie Hedger *et al.* 2000). Encouraging such a network infrastructure is therefore important. Sharing information and experiences with other developing countries can also be effective.

Availability of advanced technologies: Most of the advanced technology nowadays spreads across the world via private sector investment. Although precise figures are not available, the situation is reflected by the fact that net flows of capital to developing countries is for more than 75 per cent from private sources. At the same time official development assistance (ODA) is on a downward trend for the past decade. However, those flows are unevenly distributed: in Africa there is about $9 per capita aid for each dollar of foreign direct investment (FDI). In Asia it is the opposite with $5 FDI for each dollar of ODA (Radka *et al.* 2000). Governments do play a very important role though, because either they are directly involved in joint ventures or state-owned enterprises, or, even more important, they determine the so-called enabling conditions for private sector investment in advanced climate-friendly technologies.

Important dimensions of such an enabling environment are: first, a general macro-economic stability, good physical and telecommunications infrastructure, a good array of supporting or intermediate services in consulting, legal advice, a transparent tax structure, efficient local governance, and no widespread corruption.

Second, protection of intellectual property rights is often an important consideration for foreign investors or technology owners when deciding about providing technologies or licensing. On the other hand, too much of a restrictive attitude of technology owners to provide patent licences or otherwise (so-called restrictive business practices) in order to protect their own market or profitability can limit the spread of useful environmentally sound technologies (World Bank 1998). The international Trade Related Aspects of Intellectual Property (TRIP) agreement tries to homogenize legislation among countries, but also sets procedures that allow countries in special cases to provide so-called compulsory licences, i.e. allowing use of a patent with appropriate compensation against the will of the owner (Grubb *et al.* 2000).

Third, establishing adequate local R&D programmes in developing countries, with or without international collaboration, allows for adjustment of new technologies to the local conditions. That is often a key condition for successful diffusion of such new technologies.

Many of these developing-country actions to promote the introduction and diffusion of new climate-friendly technologies are undermined by the widespread use of export subsidies by industrialized governments. This can seriously distort the price signals and make it hard for new technologies to compete. Eliminating export subsidies for such technologies by industrialized-country governments is thus also a high priority.

Financing: Financing is a key area for improving the penetration of environment- and climate-friendly technologies. There is a clear role for public and private financing. For long-term investments, such as in transportation, energy infrastructure, or coastal protection, public finance remains central. In that context ODA is still extremely important for the least-developed countries and the declining trend needs to be reversed. Public finance also plays a role in subsidizing technologies. There is very strong evidence that existing subsidies in many cases are promoting environment- and climate-unfriendly investments while also being economically harmful (van Beers and de Moor 2001). Subsidies to promote innovative climate-friendly technologies are important to overcome the high up-front investments and to stimulate technological learning. Public finance via multinational development banks suffers from risk-averse behaviour and high project development costs putting smaller projects at a disadvantage. Greening of these large financial flows could make a major contribution (Mansley *et al.* 2000; Sathaye *et al.* 2001).

Private financing for innovative technologies is often problematic, particularly in developing countries. Commercial banks are not geared towards projects on energy efficiency or climate change mitigation, they are generally risk-averse, and they do not easily reach the community level, where small-scale credit is needed. New forms of financing have been developed such as micro-credit schemes. A successful new approach is being offered in some countries by so-called technology intermediaries.

Energy service companies, for instance, invest in a company's energy conservation if the benefits of saved energy costs are going back to them. Public–private partnerships are increasingly seen as important to help overcome the problem of lack of available capital (Mansley *et al.* 2000; Sathaye *et al.* 2001).

Training and capacity-building: An area of traditional attention is the building of human and organizational capacity in developing countries, an obvious condition for successful technology transfer. In the past, emphasis has been strongly on education and training of technical skills. More attention is needed for the full range of skills in management, organization, technical and economic evaluation, marketing, financing, etc. that are important to make innovation and implementation of new technologies a success. Public participation in decision-making and community-level involvement can be an important contribution to ensure that the needs of people are met (McKenzie Hedger *et al.* 2000).

Role of International Collaboration

It is evident that the task of pursuing a rapid diffusion of new environmentally friendly technologies throughout the world requires international collaboration. Unfortunately current mechanisms seem to be inadequate. The Climate Change Convention's provisions on technology transfer have so far only led to heated and polarized North–South debates, where the South is demanding the 'donation' of modern technologies and the North argues that technology is in the hands of the private sector, over which they have no control. The Global Environment Facility of the World Bank, United Nations Development Programme, and United Nations Environment Programme have only made a limited impact on the patterns of technology diffusion in the world, mainly because they cover only a fraction of the investment flows that go through multilateral development banks (Radka *et al.* 2000). ODA flows are decreasing. Bilateral cooperation on this issue is limited. Opportunities exist to improve the contribution of all of the mechanisms indicated to the strengthening of the diffusion and implementation of climate-friendly technologies and they should be pursued strongly. In addition, the integration and coordination of the various efforts at the country level could make a significant contribution. In most cases activities are not well integrated at present. Since the successful transfer of technologies depends on a whole range of conditions being fulfilled, neglecting some aspects can make other efforts quite ineffective. The concept of national systems of innovation (NSI) has been developed as a way of thinking about better integration of stakeholders, policy actions, research, international assistance, capacity building, access to information, enabling conditions, and all other relevant aspects of technology transfer (OECD 1999; McKenzie Hedger *et al.* 2000). NSIs would make ongoing efforts more effective and could provide focus and guidance for additional actions in line with national priorities. It can, in particular, improve the use of available international assistance, covering bilateral and multilateral activities in a coordinated manner. The Technology Co-operation Agreement Pilot Project, which the United States is implementing with a number of developing countries, is an excellent example of this approach (National Renewable Energy Laboratory 1998). The NSI approach is also applicable to developed countries and has been tried at regional level in some countries.

Coordinated action among countries is also important to address competitiveness concerns. Measures affecting internationally competing companies are often only taken when similar actions affect all competitors. This can also avoid complex debates about conflicts with international trade rules. A good example is the benefits that can be obtained from a broad international system of emissions trading (Bashmakov *et al.* 2001; Hourcade *et al.* 2001).

CONCLUSIONS

On the basis of the above the answer to the question 'Can technology save us from global climate change?' can be: 'Yes, but . . .'. The technological potential exists to avoid enough greenhouse gas emissions over the next hundred years to limit the concentrations in the atmosphere to levels in the order of 450 ppmv CO_2. Together with the effect of some other greenhouse gases this could indeed limit the risks of climate change to a low to moderate range. Nevertheless, this is not a risk-free future. Given the long time that is available for such technologies to penetrate the market there is no need for unrealistic assumptions about the penetration rates. The costs of reducing greenhouse gas emissions to meet this 450 ppmv stabilization level are also not very high, provided optimal timing of actions is applied in order to avoid premature retirement of infrastructure and installations, and no investments are made that would further lock in economies into a fossil-fuel-intensive mode. In the short term there are already many opportunities for emissions reductions that are either profitable or low-cost. For certain sectors, accompanying policies may have to be developed in order to soften adjustment to a new situation, particularly regarding the re-employment of people in different sectors of the economy.

The real problem will lie in overcoming the many political, economic, technical, social, and behavioural obstacles to putting specific measures into practice, even though these actions are technically sound and economically not significant. Coordinated action by governments is vitally needed to create the right conditions for technologies to penetrate. These actions range from making prices reflect the full social and environmental costs; taking regulatory action where markets are not properly functioning owing to behavioural or social barriers; strongly increasing the funding for and cooperation on R & D; and creating an enabling environment for the private sector to develop, implement, and transfer new climate-friendly technologies with particular emphasis on application of these new technologies in developing countries. It will not happen automatically, but it can be done.

REFERENCES

Baklid, A., and Korbol, R. (1996). *Sleipner Vest CO₂ Injection into a Shallow Underground Aquifer*, SPE 36600. Richardson, Tex.: Society of Petroleum Engineers.

Bashmakov, I., Jepma, C., Bohm, P., Gupta, S., *et al.* (2001). 'Policies, measures and instruments', in IPCC (2001c).

CADDET (1998). CADDET-Energy Efficiency—http://www.caddet-ee.org; CADDET-Renewable Energy—http://www.caddet-re.org; GREENTIE—http://www.greentie.org

Grubb, M., Ramakrishna, K., Chung, R. K., Corfee-Morlot, J., *et al.* (2000). 'International agreements and legal structures', in IPCC (2000b).

Gruebler, A. (1998). *Technology and Global Change*. Cambridge: Cambridge University Press.

Herzog, H., Adams, E., Akai, M., Alendal, G., *et al.* (2001). 'Update on the international experiment on CO₂ ocean sequestration', in P. J. Williams, R. A. Durie, P. McMullen, C. A. J. Paulson, and A. Y. Smith (eds), *Proceedings of the Fifth International Conference on Greenhouse Gas Control Technologies*, Cairns, Australia, 13–16 Aug. 2000. Collingwood: CSIAO Publishing.

Hourcade, J. C., Shukla, P., Cifuentes, L., Davis, D., *et al.* (2001). 'Global, regional and national costs and ancillary benefits of mitigation', in IPCC (2001c).

International Energy Agency (2000). *Experience Curves for Energy Technology Policy*. Paris: International Energy Agency.

International Energy Agency (2001). *IEA Energy Technology R & D Statistics: 1974–1999*, Paris: International Energy Agency. http://www.iea.org/stats/files/rd.htm

IPCC (2000a). *IPCC Special Report on Emissions Scenarios*, ed. N. Nakicenovic, J. Alcamo, G. Davis, B. De Vries, *et al.* Cambridge: Cambridge University Press.

IPCC (2000b). *IPCC Special Report on the Methodological and Technological Aspects of Technology Transfer*, ed. B. Metz, O. Davidson, J. W. Martens, S. van Rooyen, and L. van Wie. Cambridge: Cambridge University Press.

IPCC (2000c). *Land-Use, Land-Use Change and Forestry*, ed. R. T. Watson, I. R. Noble, B. Bolin, N. H. Ravindranath, *et al.* Cambridge: Cambridge University Press.

IPCC (2001a). *Climate Change 2001: The Scientific Basis; Contribution of Working Group I to the Third Assessment Report of the IPCC*, ed. J. Houghton, Y. Ding, D. J. Griggs, M. Noguer, *et al.* Cambridge: Cambridge University Press.

IPCC (2001b). *Climate Change 2001: Impacts, Adaptation and Vulnerability; Contribution of Working Group II to the IPCC Third Assessment Report*, ed. J. J. McCarthy, O. F. Canziani, N. A. Leary, D. J. Dokken, and K. S. White. Cambridge: Cambridge University Press.

IPCC (2001c). *Climate Change 2001: Mitigation; Contribution of Working Group III to the IPCC Third Assessment Report*, ed. B. Metz, O. Davidson, R. Swart, and J. Pan. Cambridge: Cambridge University Press.

IPCC (2002). *Climate Change 2001: Synthesis Report*, ed. R. T. Watson. Cambridge: Cambridge University Press.

Jochem, E., Sathaye, J., and Bouille, D. (eds) (2001). *Society, Behaviour and Climate Change Mitigation*. Dordrecht: Kluwer.

Kauppi, P., Sedjo, R., Apps, M., Cerri, C., *et al.* (2001). 'Technological and economic potential of options to enhance, maintain and manage biological carbon reservoirs and geo-engineering', in IPCC (2001c).

McKenzie Hedger, M., Martinot, E., Onchan, T., Ahuja, D., *et al.* (2000). 'Enabling environment for technology transfer', in IPCC (2000b).

Mansley, M., Martinot, E., Ahuja, D., Chantanakome, W., *et al.* (2000). 'Financing and partnerships for technology transfer', in IPCC (2000b).

Mitchell, J. V., and Dolun, M. (2001). *The Fuel Tax Protests in Europe 2000–2001*. London: Royal Institute of International Affairs.

Moomaw, W. M., Moreira, J. R., Blok, K., Greene, D. L., *et al.* (2001). 'Technological and economic potential of greenhouse gas emissions reductions', in IPCC (2001c).

Morita, T., Nakicenovic, N., and Robinson, J. (2000). 'Overview of mitigation scenarios for global climate stabilization based on new IPCC emission scenarios (*SRES*)', *Environmental Economics and Policy Studies*, 3 (2): 65–88.

Morita, T., Robinson, J., Adegbulugbe, A., Alcamo, J., *et al.* (2001). 'Greenhouse gas emission mitigation scenarios and implications', in IPCC (2001c).

National Renewable Energy Laboratory (1998). *Technology Co-operation Agreement Pilot Project: Development Friendly Greenhouse Gas Reductions. Status Report.* Golden, Colo.: NREL.

OECD (1999). *Managing National Innovation Systems*. Paris: OECD.

Prentice, I. C., Farquhar, G. D., Fasham, M. J. R., Goulden, M. L., *et al.* (2001). 'The carbon cycle and atmospheric carbon dioxide', in IPCC (2001a).

Radka, M., Aloisi de Larderel, J., Brew-Hammond, J. P. A., Xu, H. (2000). 'Trends in technology transfer: financial resource flows', in IPCC (2000b).

Ramaswamy, V., Boucher, O., Haigh, J., Hauglustaine, D., *et al.* (2001). 'Radiative forcing of climate change', in IPCC (2001a).

Rotmans, J., Kemp, R., van Asselt, M. B. A., Geels, F., *et al.* (2001). *Transitions and Transition Management: The Case for a Low Emission Energy Supply*. ICIS working paper Io1-Eoo1. Maastricht: International Centre for Integrative Studies.

Sathaye, J., Bouille, D., Biswas, D., Crabbe, P., *et al.* (2001). 'Barriers, opportunities and market potential of technologies and practices', in IPCC (2001c).

Stevens, S. H., Spector, D., and Riemer, P. (1999). 'Enhanced coal bed methane recovery by use of CO_2', *Journal of Petroleum Technology*, 62: 62.

Tóth, F., Mwandosya, M., Carraro, C., Christensen, J., *et al.* (2001). 'Decision-making frameworks', in IPCC (2001c).

UNFCCC (The United Nations Framework Convention on Climate Change) (1992). http://www.unfccc.int/resources

van Beers, C., and Moor, A. de (2001). *Public Subsidies and Policy Failures*. Cheltenham: Edward Elgar.

van Vuuren, D. P., and de Vries, H. J. M. (2001). 'Mitigation scenarios in a world oriented at sustainable development: the role of technology, efficiency and timing', *Climate Policy*, 1: 189–210.

Williams, R. H., Bunn, M., Consonni, S., Gunter, W., *et al.* (2000). 'Advanced fossil technologies', in J. Goldemberg (ed.), *World Energy Assessment: Energy and the Challenge of Sustainability*. New York: United Nations Development Programme, Department of Economic and Social Affairs, and World Energy Council.

Wilson, M., Moburg, R., Stewart, B., and Thambimuthu, K. (2001). 'CO_2 sequestration on oil reservoirs—a monitoring and research opportunity', in P. J. Williams, R. A. Durie, P. McMullen, C. A. J. Paulson, and A. Y. Smith (eds), *Proceedings of the Fifth International Conference on Greenhouse Gas Control Technologies*, Cairns, Australia, 13–16 Aug. 2000. Collingwood: CSIAO Publishing.

World Bank (1998). *World Development Report 1998*. Washington: World Bank.

5

The Role of Corporate Leadership

John Browne

THE title of this volume is *Managing the Earth*, which implies that the Earth can be managed and that a series of diverse issues can be reduced to a set of relationships which are susceptible to the discipline of management. That is a bold and brave assumption. I do not know if that is true, but I think testing that assumption is the right approach because it forces us to consider all the difficult questions on the basis of facts and logic and it avoids the risk of important arguments being lost in clouds of rhetoric and emotion. Emotion on environmental issues may be very understandable. People worry and feel angry but, if we are ever going to make progress, those concerns need to be subjected to rational analysis.

So how would one start to manage the Earth? My approach is to apply the two basic principles of management which apply in any situation: reality and purpose. Let us start with reality and with some facts.

The first reality is the growth of the world's population. Up by 160 per cent in my lifetime from 2.5 billion to 6 billion and rising now by 90 million a year—which means that enough people are born every nine hours to populate a city the size of Oxford. Every one of those people needs energy.

To me, the instinct not just to survive, but to strive for a better life and to try to improve your living standards, is fundamental to human nature. To do that—to have food, to have a home, to have light and heat and mobility, even at the most basic level—requires energy. Energy is what keeps people alive, and what allows them to live with some element of dignity. That is the second reality.

The third reality is that for the foreseeable future the bulk of the energy that the world needs will come from hydrocarbons and particularly from oil and gas. That is not an assertion on behalf of the oil and gas industry: it is a simple fact.

Oil and gas currently supply 65 per cent of the world's daily needs[1] and over the next two decades at least that figure will increase. Nuclear power remains expensive, particularly if you include the full costs of disposing of the waste products. Coal is still important in some areas but it carries enormous environmental costs in terms of emissions. In the main energy-using activities—in transportation, industry, domestic supply, and power generation—the most convenient fuels in every sense are oil and gas.

Today we use over 75 million barrels per day (mb/d) of oil and 220 billion cubic feet per day (bcf/d) of gas. By the end of this decade on fairly conservative assumptions about economic growth the world will use more than 90 mb/d of oil and 280 bcf/d of gas, and those figures will still be rising.

What about renewables and hydrogen? Are they the answer? One day they may well be part of the answer, and given the scale of the need for energy I believe it is prudent to start developing them as viable additional sources. This will require coordinated action by governments, consumers, and companies in the private sector. We need to achieve the necessary economies of scale to ensure that early investment is worth while, and that the resulting energy prices are attractive to consumers.

Currently renewable energy, excluding the large-scale hydro plants, represents only 1 per cent of primary energy production worldwide, and 2 per cent of global electricity-generating capacity. At the moment renewable energy is dominated by small-scale hydroelectric projects and biomass, but in the future other sources will play a bigger role in fuelling at least a small part of the growth in electricity consumption over the next twenty years—which some predict to be as much as 80 per cent.

BP and others are doing significant work on the technology of photovoltaics—solar power. We are also developing the use of hydrogen as the ultimate clean source of fuel for vehicles. Hydrogen would not generate any tail pipe emissions—except for water of drinking-quality standard. We have partnerships with Ford and GM and Daimler Chrysler to promote the use of experimental hydrogen-powered fuel cell vehicles in London, Sacramento, Sydney, and Beijing, and a project with BMW to demonstrate the benefits of using hydrogen in conventional internal combustion vehicles. Within the next two years we will begin supplying hydrogen from our refinery at Kwinana in Australia and we aim to have one of the world's first hydrogen retail stations.

Those are very exciting developments, and of course many other companies are also investing and making progress in this area. However, these must currently be recognized as experiments with future potential. They

---

[1] The energy statistics in this paper are drawn from the BP Statistical Review of World Energy (BP 2001).

should not take our eyes off the current reality, which is that the world will continue to rely on oil and gas for a very long time to come.

Supplies are available to meet that demand. On the best estimates, the world has found and produced around 800 billion barrels of oil and natural gas liquids. The remaining reserves are around 950 billion barrels, much of which is in the Middle East. However, we believe that there are another 500 billion barrels of additional supplies which can come from new discoveries and improvements in the recovery rates in existing fields. Outside the Middle East quite a large proportion of the oil still to be found and developed is in the deep water beyond the continental shelf, but technology is making the deep water accessible in ways that seemed impossible only ten or fifteen years ago. In terms of natural gas the figures are much larger: only some 20 per cent of estimated world total natural gas supplies have so far been found and produced, and huge supplies remain in many different parts of the world. So, leaving aside politics, there need be no shortage of oil and gas.

A fourth reality is that the growing consumption of oil and gas poses an environmental challenge. At one level there is the challenge of low-level pollution and poor air quality—particularly in the cities. In the longer term there is the risk of climate change through global warming.

Of course, the science of climate change presently remains unproven, but no one reading the latest scientific reports published by the Intergovernmental Panel on Climate Change in 2000–1 could ignore the mounting evidence of a link between human activity and the world's climate, or the implications (IPCC 2001*a*). The latest report concentrates on the likely consequences (IPCC 2001*b*).

Rising water levels, which put particular communities at risk, and floods in some areas are matched by drought and a decline in long-term water levels, which risk disrupting agricultural systems and adding to the problems of some of the world's poorest areas. To put that in another perspective, a recent report by the insurance industry has estimated that the damage could cost the global economy over $300 billion a year (Berz 2001).

The four points that I have set out and discussed above represent the reality. If management of the earth is to have any effect, it has to begin with reality—the whole reality. Of course, that is not a comfortable analysis, for the world or for the industry of which I am part, because it implies that the status quo will not continue, and that some change is necessary.

If that *is* the reality, what about the other element of management: purpose? The purpose of business is to do business—to make profitable investments, and to deliver competitive returns on behalf of our shareholders. That is a very simple purpose—clear and limited. Businesses do not exist to run the world, and have no legitimacy to do so. We cannot take decisions for people—because we have no authority. No one elected us.

Clearly, companies are part of society. We fulfil a specific role on behalf of society—creating wealth and meeting needs. But we're not responsible for society. We have a single simple purpose but, of course, that purpose cannot be fulfilled in isolation. No company, however big, is sufficient unto itself.

A company's ability to fulfil its purpose depends on the decisions and choices made every day by all the people with whom they do business—governments, other companies, staff, and, in the case of BP, the 10 million or so people who buy things from us around the world every day. Our ability to do business is determined by our capacity to meet the needs of those people, and those needs in turn are being set by the reality I have described.

I believe the challenge—the business challenge—is to transcend the sharp trade-off implied by this analysis. Put in the simplest terms, the trade-off is that the world has a choice: economic growth, fuelled by increasing energy consumption, or a clean environment. We can have one or the other, but not both.

If that is the trade-off, it is unacceptable. People want both. And I believe there is a huge commercial prize for those who can offer better choices which transcend the trade-off. And that was the objective we set ourselves in BP four years ago, when I gave a speech at Stanford which said that some action, of a precautionary nature, particularly on climate change, was essential. If no precautionary steps were taken there would be a risk that drastic action would become necessary with the risk of serious disruption to the world's economy.

Four years on this seems the right moment to review what we aimed at against the reality of the track record. This is a good moment to see if the aspiration is justified. I will talk about BP, but I think it is important to make clear that there are many people looking for ways to transcend the trade-off, and this is not an isolated story. Many of things we have achieved could not have been done without the help of our partners and contractors and suppliers. So the focus simply reflects the fact that this is the only story I am really qualified to tell.

We set ourselves a number of goals. First, the objective of reducing our own emissions of carbon dioxide by an absolute amount of 10 per cent from a 1990 base, while growing the company. We have made good progress. By the end of last year we had delivered a reduction of 5 per cent and we can identify at least another 5 per cent which is deliverable over the next three years.

The progress we have made has not come from the use of a single magic bullet—rather dozens and dozens of initiatives, most of them undertaken at local level by our business unit leaders and their teams which cumulatively have delivered material progress. Reducing flaring; tightening the control of emissions from our refineries; reducing our own energy use: they all count. Let me quote just two examples.

In our acetic acid production unit in Hull we found that we were producing less acetic acid for the volume of carbon monoxide being used than should have been the case. We traced the problem to a leak from the compressors. A new advanced sealing process was created by the engineers and the result was an increase in production of acetic acid of 20,000 tonnes a year, and reduction of 15,000 tonnes year in the emissions of carbon dioxide—a reduction of 98 per cent with all the benefits that brings in terms of both atmospheric pollution and health and safety in the plant.

A second example: in the United States venting accounts for almost 60 per cent of all the methane which is emitted by our gas business in the western states. To reduce that, we have a project under way to control emissions from all wells in the areas of the Greater Green River and the San Juan basin. This is not an easy project, but we have now found that with an investment of $1.4 million we can save more than 20,000 tonnes of methane a year, which has the greenhouse effect of nearly half a million tonnes of carbon dioxide. Moreover, the methane can now be sold—turning what began as an environmental project into something that is also good profitable business.

Reducing emissions was the first goal in our company. The second was to demonstrate that we could create an internal trading system which would allow us to meet our target at the lowest possible cost by allocating the resources to the places where they would have most effect, i.e. meeting the target, not by requiring everyone to cut emissions by 10 per cent, but by putting a monetary value on each unit and encouraging the whole internal team to cooperate in delivering the target at the lowest cost.

Our emissions trading system began in 1999, and has been working across the whole company since January 2000. In the first year 2.7 million tonnes of carbon dioxide were traded at an average price of $7.6 per tonne. The system now covers every single operation we have around the world. We have learned a lot, and now we are ready to develop the system and perhaps to bring in third parties who will allow us both to have a greater impact and to reduce unit costs. This approach has encouraged our business units to look for innovative, cost-effective ways of meeting the target.

Our third goal was to make a positive contribution to the problem of air quality. We have developed a series of clean fuels: gasoline and diesel without lead, sulphur, or benzene. That programme was launched in January 1999. By the end of last year it had reached fifty-nine cities worldwide. By the end of this year those products will be available in ninety cities, including many where there are serious and persistent air quality problems. We will continue to extend the reach of that pro-gramme and to work on the technology so that we can provide even cleaner fuels. And we will work with the car companies and others who share our view that the wonderful achievement of individual mobility— one of the great advances of the last hundred years—does not have to be compromised and tarnished by pollution.

The fourth goal was to make a contribution to the shift in the energy mix. That mix is always changing. When I was born, the dominant fuel in this country was coal—which is why so many beautiful buildings in this city and elsewhere were covered in grime and dust. There has been a shift, here and worldwide, from coal to oil, and now there is a further shift going on in favour of natural gas. We are part of that process.

Five years ago natural gas represented no more than 15 per cent of our business. Now it is 40 per cent and still rising, not just in the developed world—where we are the largest producer in the United States—but in areas where energy demand is growing most rapidly and where the choice of supply to meet that demand will have huge environmental con-sequences. In China, for instance, we intend to be a major supplier of nat-ural gas—helping the country to escape from a dependence on coal.

All the things I have mentioned are 'work in progress'. It is far too early to celebrate victory. But I believe that we have demonstrated that it is pos-sible to combine economic growth with a progressive improvement in the environmental impact of that growth. There is much more to do, but on the basis of the track record so far there is room for cautious optimism—which is all too rare a commodity in the environmental debate which is still too often reduced to slogans and denial.

Of course, just as no company stands alone, so no single company on its own can do everything that is necessary around the world. Progress depends on the cooperation of different elements of society if there is to be a material change. This is where pessimism seems to have taken over the debate. There is a sense that the limited progress made since Kyoto indic-ates that cooperation is impossible. I disagree. Here again I think there are grounds for cautious optimism, and not only because Kyoto was just one stage in the process of discussion, and not an end point.

I think we can usefully compare the debate on the long-term issue of climate change to the debates on free trade from the initial establishment of the General Agreement on Tariffs and Trade (GATT) or the debate on dis-armament which has been running since the early treaties of the post-war period. Both took half a century, and both are still incomplete—but through successive steps they both made great progress. And that I think is how this debate will proceed—not through one step to a perfect solution, but through one step to another step.

Linked to that, the second reason for optimism is the technical progress that has been made in the academic world, and in the business community: progress in understanding the problems in detail and in offering answers. An important part of that progress is the success of the trading efforts that have been initiated by my company, and by others, and the potential which trading offers to remove the fear that the cost of dealing with this problem was unmanageable and that any action would cost enormous numbers of jobs.

One of the most interesting studies published in the last few months was the report from the Pew Centre[2] which showed that the cost of reducing emissions to the Kyoto target levels could be cut from an estimated $57 billion a year to less than $9 billion if a global trading system were used to allocate resources, and by $20 billion to $37 billion even if the trading system just embraced the countries in Annex 1 of the Kyoto Protocol. I imagine that there will still be significant savings even if the trading system is established just within individual countries.

The fourth reason for optimism is the approach of some key governments. I was particularly struck recently by listening to a talk given to the Business Council, a group of the most senior business leaders in the United States, by the new administrator of the Environmental Protection Agency, Governor Whitman. Her crucial points were, first, that the prime, overriding objective of the new US administration in the environmental area is to reduce pollutants—in particular, carbon dioxide and sulphur—and to use trading to help achieve that. Second, the administration takes very seriously and wants to respond positively to the latest IPCC findings—with practical action that moves beyond the Kyoto discussions. Third, the administration is in the business of setting objectives, not of prescribing the means—which I interpret (and I stress this is my comment not hers) as desire not to repeat the sort of mistakes made in the prescription of the precise chemical formulation of gasoline. Fourth, she insisted that while there is an inclination to use carrots to encourage a positive and creative response from business, the stick has not been thrown away. I take those comments very seriously, because I think they build on the foundations laid previously in a very constructive way.

Other governments have also taken steps which set objectives and establish incentives for progressive change: in the United Kingdom, for instance, with the incentives for motorists to switch to cleaner fuels, and in China through encouragement of new sources of energy supply again focused on the use of gas. So I see grounds for cautious optimism—supported by the advances in technology and by the practical use of public power.

[2] www.pewclimate.org

I emphasize one other major factor—globalization. I know that in some circles globalization is much abused and regarded as the source of every problem we face. Globalization is a complex and incomplete process but I do not think that anyone who cares about the environment could seriously regard its impact as negative—quite the reverse. Most of the advances I have discussed flow from the spread of knowledge which is the unique characteristic of globalization: knowledge of the challenge and knowledge of the potential solutions. This is knowledge which is transmitted not just in the public domain—in the media—but also through particular networks, notably within the academic world, and within companies.

Almost all the steps we have taken in BP, for instance, to reduce emissions began with a particular advance in a particular place. They have spread and become common practice because we have networks within the company—technical specialists sharing experience and ideas on a world-wide basis, and they are enabled to spread quickly because of the advances in communications. When you see it in action, that is a magical process, and it is exemplary of what is becoming possible as globalization creates a single community of knowledge. That is just what is needed to make progress.

So, to come back to the challenge of the title, there are sufficient grounds for cautious optimism, based on what appear to be practical advances. If we apply the principles of management—reality and purpose—we can confront the problems and help to create and spread the solutions. The answer to the environmental challenge of economic growth is neither denial nor retreat. The answer is not to say that there is no problem or that the problem is so intractable that we have to put a halt to the whole process of economic advance. The answer lies in allowing and encouraging the process of economic development to resolve its own contradictions. The paradox is that the answer to the problems created by development lies in more development. That has been the story of human progress so far, and I believe we are now seeing that story rewritten again.

REFERENCES

Berz, G. (2001). 'Insuring against catastrophe'. *Our Planet* (Feb.), 1–6.

BP (2001). *BP Statistical Review of World Energy 2000*. London: BP. Updated version available at http://www.bp.com/centres/energy

IPCC (2001a). *Climate Change 2001: The Scientific Basis; Contribution of Working Group I to the Third Assessment Report of the IPCC*, ed. J. Houghton, Y. Ding, D. J. Griggs, M. Noguer, *et al.* Cambridge: Cambridge University Press.

IPCC (2001*b*). *Climate Change 2001: Impacts, Adaptation and Vulnerability; Contribution of Working Group II to the IPCC Third Assessment Report*, ed. J. J. McCarthy, O. F. Canziani, N. A. Leary, D. J. Dokken, and K. S. White. Cambridge: Cambridge University Press.

6

Who Governs a Sustainable World? The Role of International Courts and Tribunals

Philippe Sands

INTRODUCTION

IN 1992, at the United Nations Conference on Environment and Development (UNCED), states adopted the Rio Declaration on Environment and Development. Principle 27 of the Rio Declaration states: 'States and people shall co operate in good faith and in a spirit of partnership in the fulfilment of the principles embodied in this Declaration and in the further development of international law in the field of sustainable development' (United Nations 1992). The language of Principle 27 is premissed on the view that even at the time of its adoption there existed a body of 'international law in the field of sustainable development'. However, Principle 27 does not indicate the content of that law, in particular whether it is procedural or substantive or both, or where its content may be identified.

Shortly after the adoption of the Rio Declaration a group of independent legal scholars and practitioners sought to identify its content, on the basis of a review of legal and policy instruments and the international practice of states (which was then, and remains now, somewhat limited). The group, the Foundation for International Environmental Law and Development, concluded that

the concept of 'sustainable development' is now established in international law, even if its meaning and effect are uncertain. It is a legal term which refers to processes, principles and objectives, as well as to a large body of international agreements on environmental, economic and civil and political rights. (Foundation for International Environmental Law and Development 1993: 1)

The author is indebted to Noémi Byrd for her help with the preparation of this paper.

In 1994 I wrote a piece (Sands 1994) in the *British Yearbook of International Law* in which I sought to review what these processes, principles, and objectives might be. My conclusion was that 'international law in the field of sustainable development' coalesced around

a broad umbrella accommodating the specialised fields of international law which aim to promote economic development, environmental protection and respect for civil and political rights. It is not independent and free-standing of principles and rules, and it is still emerging. As such, it is not coherent or comprehensive, nor is it free from ambiguity or inconsistency.... The significance of the UNCED process is not that it has given rise to new principles, rules or institutional arrangements. Rather, it endorses on behalf of the whole of the international community (states, international institutions, non-governmental actors) an approach requiring existing principles, rules and institutional arrangements to be treated in an integrated manner. (Sands 1994: 379)

That was 1994. Since then there have been numerous developments relating to the definition and application of the concept of sustainable development. The term has been incorporated into many national laws. In Britain, for example, Section 4 of the Environment Act of England and Wales 1995 says that 'it shall be the principal aim of the [Environment] Agency . . . in discharging its functions so to protect or enhance the environment . . . as to make the contribution towards attaining the objective of sustainable development'. Other national provisions provide guidance on particular areas or activities.[1] The term has also been incorporated into treaties, resolutions, and other acts of international organizations, and is regularly invoked to support all manner of positions which states and other actors—corporations and non-governmental organizations in particular—seek to justify, both at national and international levels. A careful reader of each of these national and international instruments will divine that in none of these acts is the term defined with any degree of precision, or at all. Certainly no definition I have seen lends the concept to clear and practical application. But issues of practical application abound.

This lack of clarity raises an issue of governance. Assuming that the concept of sustainable development has some sort of legal status and meaning, and if it is true that it remains generally undefined in the instruments to which I have referred, or defined differently in different contexts, and if it

---

[1] The Environment Agency, for example, has adopted a Policy Guidance, which provides that 'In terms of the Agency's duty relating to sustainable development, policies or practices that promote unsustainable long-term options are not acceptable. Promotion of the use or proliferation of cesspools is therefore unacceptable, because cesspools require regular maintenance and emptying in order to operate effectively'; cited in *R (On the Application of Anglian Water Services Ltd) v Environment Agency*, Queen's Bench Division, Judgment of 24 October 2000, Tomlinson J.

is the case that is intended to have practical application, then which person or body is to decide whether an activity is consistent with sustainable development—what the term means in any given case? The traditional options include the legislative branch, the executive or administrative branch, and the judiciary. Given the absence of clear legislative guidance, and the limited role of an international executive branch, in this paper I want to explore the role of the judiciary—in particular the international judiciary—in giving meaning to the concept of sustainable development. My thesis, in summary, is that the inherent ambiguity of the concept of sustainable development implies a greater role for the international judiciary. One of the notable developments of the late twentieth century is the growth of the international judiciary, a development that has occurred almost without debate or discussion. A generation ago the International Court of Justice in The Hague was out of sight and almost out of mind as it processed one or two cases a year. In the 1950s two new international bodies were created, again without a great deal of public interest or comment: the European Court of Human Rights, based in Strasbourg, charged with overseeing the implementation by states of the European Convention on Human Rights, and the European Court of Justice, based in Luxembourg, with responsibility for European Community law. Other regional bodies followed in Africa and the Americas. But it was in the 1990s that the real transformation took place. The UN Security Council created international criminal tribunals for the former Yugoslavia and Rwanda, states created an enormously powerful judicial organ to resolve trade disputes under the auspices of the World Trade Organization, and in the summer of 1998, 160 states adopted the Statute of a new International Criminal Court (Sands *et al.* 1999).

When I wrote the piece for the *British Yearbook of International Law*, in 1994, the term 'sustainable development' had not been the subject of international judicial consideration. 'Sustainable development' has now been invoked before bodies charged with resolving international disputes, including the International Court of Justice (ICJ) and the Appellate Body of the World Trade Organization. In this paper I want to consider what the jurisprudence of those two bodies has added to our understanding of 'sustainable development', and what it tells us about the role of international courts in our emerging 'system' of international governance.

THE INTERNATIONAL COURT OF JUSTICE

The concept of 'sustainable development' received its first thorough airing in the case concerning the Gabcíkovo–Nagymaros project between

Hungary and Slovakia before the ICJ (ICJ Reports 1997: 7ff.).[2] The case concerned a dispute over whether or not to build two barrages on the Danube shared by Hungary and Czechoslovakia. In 1977, by treaty, the two countries had agreed to build two barrages which would then be jointly operated. The 1977 Treaty envisaged the diversion of waters from the Danube, where it was a boundary river, onto Czechoslovak territory, and the operation of a dual system of barrages by 'peak power' (rather than 'run of the river' mode). Construction began and proceeded more slowly than had been originally envisaged. In the mid-1980s political opposition in Hungary focused on the environmental aspects of the barrage as a means of achieving broader political change. In May 1989 great public pressure led Hungary to suspend work on large parts of the project. The two countries sought to reach an agreement on how to proceed. Both were intransigent and committed to different approaches. Czechoslovakia took the view that the barrages posed no serious threat to the environment; Hungary was certain they would lead to significant environmental harm to water supplies and to biodiversity. For Czechoslovakia construction constituted 'sustainable development', for Hungary only termination of the whole project could be consistent with 'sustainable development'. In the absence of an agreed resolution of the problem, and in the face of Hungary's refusal to continue work on the project, in 1991 Czechoslovakia proceeded unilaterally to implement what it termed a 'provisional solution' (referred to as Variant C), comprising a single barrage on the Czechoslovakian side, but requiring the diversion of some 80 per cent of the shared water onto its territory. It argued that this was justified by the 1977 Treaty, which, in effect, gave it rights over that amount of water for the purposes of operating a barrage on its side. As Variant C proceeded in late 1991 and early 1992 Hungary took the view that it had no option but to terminate the 1977 Treaty, which apparently provided the sole basis upon which Czechoslovakia claimed to be able to proceed to its unilateral and provisional solution. In May 1992 Hungary purported to terminate the 1977 Treaty, a complicated situation which was made no easier when, in January 1993, Czechoslovakia split into two countries, with the Czech Republic and Slovakia agreeing between themselves that Slovakia would succeed to ownership of the Czechoslovak part of the project. In the meantime, in October 1992 Czechoslovakia had dammed the Danube and diverted over 80 per cent of the waters of the Danube into a bypass canal on Slovak territory. In April 1993, largely under the pressure of the Commission of the

---

[2] The Court had previously referred to Principle 24 of the Rio Declaration on Environment and Development in its Advisory Opinion on the Legality of the Threat or Use of Nuclear Weapons (ICJ Reports 1996: 242).

European Communities, Hungary and Slovakia agreed to refer the matter to the ICJ. The Court was presented with an opportunity to address a wide range of international legal issues, including the law of treaties, the law of state responsibility, the law of international watercourses, the law of the environment, and the interrelation of these areas. Against this background, of course, the concept of 'sustainable development' had also emerged into international legal discourse.

The Court was specifically asked to address three questions posed by the parties. What did it rule? First, it found on the facts that Hungary was not entitled in 1989 to suspend or terminate—on environmental grounds—work on the joint project. Second, it ruled that Czechoslovakia (and subsequently Slovakia) was not entitled to operate from October 1992 a unilateral solution diverting the Danube without the agreement of Hungary (although it ruled that construction prior to operation was not unlawful). Third, the Court went on to say that Hungary was not entitled in May 1992 to terminate the 1977 Treaty, which remains in force to this day. As to the future, the Court indicated the basis for cooperation and agreement which it hoped the parties might pursue, suggesting that the preservation of the status quo—one barrage not two, jointly operated, no peak power (that is to say, the barrage would operate on run-of-the-river mode)—would be an appropriate solution, in effect rewriting the 1977 Treaty. It was in relation to these future arrangements that the majority of the Court invoked the 'concept of sustainable development' to suggest a way forward. Specifically, it said this:

Throughout the ages, mankind has, for economic and other reasons, constantly interfered with nature. In the past this was often done without consideration of the effects upon the environment. Owing to new scientific insights and to a growing awareness of the risks for mankind—for present and future generations—of pursuit of such interventions at an unconsidered and unabated pace, new norms and standards have been developed, set forth in a great number of instruments during the last two decades. Such new norms have to be taken into consideration, and such new standards given proper weight, not only when States contemplate new activities, but also when continuing with activities begun in the past. This need to reconcile economic development with protection of the environment is aptly expressed in the concept of sustainable development. (ICJ Reports 1997: 78, para. 140)

The Court followed this by concluding, in the same paragraph of the judgment, that

For the purposes of the present case, this means that the Parties together should look afresh at the effects on the environment of the operation of the Gabcíkovo power plant. In particular they must find a satisfactory solution for the volume of water to be released into the old bed of the Danube and into the side-arms on both sides of the river.

At least three aspects of what the Court said are to be noted. First, the fact that it invokes 'sustainable development' at all indicates that the term has a legal function, and some sort of status in international law. Second, 'sustainable development' is a 'concept' and not a principle or a rule. And third, as a 'concept' it has both a procedural–temporal aspect (obliging the parties to 'look afresh' at the environmental consequences of the operation of the plant) and a substantive aspect (the obligation of result to ensure that a 'satisfactory volume of water' be released from the bypass canal into the main river and its original side-arms). The significance of the Court's invocation of sustainable development has been widely noted, including in an important contribution by Lowe noting the 'great interest' of the Court's approach, which is 'likely to prove to be of enormous influence' (Lowe 1999: 19).

The Court does not, however, indicate the content of the procedural–temporal requirement (for example, does this require a formal or informal environmental impact assessment? and if so, according to what standards?) or the factors for determining whether the volume of water flowing in the Danube would be said to be satisfactory. Paragraph 140 is cryptic, to say the least. During the course of written arguments both sides had invoked 'sustainable development' to justify their positions.[3] The pleadings will repay a careful study, since they reflect the inherent malleability and uncertainty of the term. Hungary invoked 'sustainable development' to justify its view that there should be no barrages, whereas for Slovakia the 'concept' justified the opposite conclusion, namely that 'sustainable development' could only be achieved if both barrages envisaged by the 1977 Treaty were constructed. It might be said that Hungary focused on the environmental aspect of the concept while Slovakia focused on its 'developmental' elements. For its part the Court invokes the concept to achieve an accommodation of views and values while leaving to the parties the task of fleshing out the harder practical consequences. The Court appears to use the concept to build a bridge, justifying a conclusion other than that which would tend to flow directly from its earlier reasoning and conclusions, namely that with its finding that the 1977 Treaty remained in force Hungary ought logically to be required to construct the second barrage at Nagymaros. Or, as two other authors have

---

[3] See e.g. Slovakia: 'It is clear from both the letter and the spirit of these principles that the overarching policy of the international community is that environmental concerns are not directed to frustrate efforts to achieve social and economic development, but that development should proceed in a way that is environmentally sustainable. Slovakia submits that these have been, and are today, the very policies on which the G/N Project is based' (Counter-Memorial, para. 9.56). In reply, Hungary takes the opposite view to support its argument that the G/N Project is unlawful: 'Well-established . . . operational concepts like "sustainable development" . . . help define, in particular cases, the basis upon which to assess the legality of actions such as the unilateral diversion of the Danube by Czechoslovakia and its continuation by Slovakia' (Hungarian Reply, para. 3.51).

put it: 'What is perhaps more remarkable, however, is that the Court, despite its endorsement of a treaty regime that smacked of unsustainability, went on to invoke sustainable development in order to miraculously salvage something from a sinking ship' (Stec and Eckstein 1997: 47).

To be clear, the Court did not rely exclusively on 'sustainable development' to justify this conclusion, having found as a matter of fact that Slovakia itself had conceded that no second barrage was now necessary: 'Equally, the Court cannot ignore the fact that, not only has Nagymaros not been built, but that, with the effective discarding by both parties of peak power operation, there is no longer any point in building it' (ICJ Reports 1997: 7). 'Sustainable development' was used to fortify that conclusion and provide some guidance regarding its consequences. Beyond paragraph 140 the Court provided no further assistance concerning the status of 'sustainable development' in international law, or its practical consequences, beyond the fact that it was to fulfil a function of integrating the potentially competitive societal objectives of environment and development. Perhaps some assistance on what the Court might have had in mind may be gleaned from the Separate Opinion of Judge Weeramantry, who joined in the majority judgment, and whose hand almost certainly guided the drafting of paragraph 140. According to Judge Weeramantry the 'principle' of sustainable development fulfilled a harmonizing and reconciling function, requiring development and environment to be treated in a balanced way to avoid 'a state of normative anarchy', and is 'a part of modern international law by reason not only of its inescapable logical necessity, but also by reason of its wide and general acceptance by the global community' (ICJ Reports 1997: 7).

These words provide some illumination of the place which 'sustainable development' may have in the international legal order, but does not indicate with any degree of precision how reconciliation or harmonization are to be achieved, or how, on the facts of this case, one barrage rather than two might better achieve the objective of 'sustainable development'. This of course is not a criticism, but rather a comment on the difficulties posed for the judicial function of measuring and then balancing competing objectives. In this sense the term 'sustainable development' appears useful as a means of bridging two views without necessarily having to provide close reasoning as to method or outcome.

WORLD TRADE ORGANIZATION APPELLATE BODY

By way of contrast, let us turn now to the approach of the Appellate Body of the World Trade Organization (WTO) in the subsequent case concerning the import prohibition imposed by the United States on Certain Shrimp and Shrimp Products from India, Malaysia, Pakistan, and Thailand, on the grounds that they were harvested in a manner which adversely affected endangered sea turtles (WTO 1998*a*; *International Legal Materials* 1999). The WTO has, of course, attracted a high degree of notoriety, culminating in the events at Seattle in November 1999 which wrecked the proposed Millenium Round of the WTO. The *Shrimp Turtle* case was a major catalyst for those events.

In 1987 the United States had issued regulations (pursuant to its 1973 Endangered Species Act) requiring all United States registered shrimp trawl vessels to use approved turtle excluder devices (TEDs) in specified areas where there was a significant mortality of sea turtles in shrimp harvesting. TEDs allowed for shrimp to be harvested without harming other species, including sea turtles. The United States regulations became fully effective in 1990, and were subsequently modified to require the general use of approved TEDs at all times and in all areas where there was a likelihood that shrimp trawling would interact with sea turtles. In 1989 the United States enacted Section 609 of Public Law 101–162, which addressed the importation of certain shrimp and shrimp products. Section 609 required the United States Secretary of State to negotiate bilateral or multilateral agreements with other nations for the protection and conservation of sea turtles. Section 609(b)(1) imposed (not later than 1 May 1991) an import ban on shrimp harvested with the commercial fishing technology which may adversely affect sea turtles. Further regulatory guidelines were adopted in 1991, 1992, and 1996, governing *inter alia* annual certifications to be provided by harvesting nations.

In broad terms, certification was to be granted only to those harvesting nations which provided documentary evidence of the adoption of a regulatory programme to protect sea turtles in the course of shrimp trawling. Such a regulatory programme had to be comparable to the programme of the United States, with an average rate of incidental taking of sea turtles by their vessels which should be comparable to that of the United States vessels. The 1996 guidelines further required that all shrimp imported into the United States had to be accompanied by a shrimp exporter's declaration attesting that the shrimp was harvested either in the waters of the nation certified under Section 609, or under conditions that did not adversely affect sea turtles, including through the use of TEDs. From a WTO perspective the difficulty was that the United States was, in effect, applying its conservation laws extraterritorially to activities carried out within—or subject to the jurisdiction of—third states. This, of course, raises a big issue of general international law, namely the circumstances (if any) in which a state may apply its own conservation measures to activities taking place outside its territory or jurisdiction, including by non-nationals.

The United States sought to justify its actions on the grounds that the sea turtles it was seeking to protect were recognized in international law as being endangered. The United States legislation was challenged by India, Malaysia, Pakistan, and Thailand. At first instance a WTO Panel concluded that the import ban applied on the basis of Section 609 was not consistent with Article XI.1 of GATT 1994 and could not be justified under Article XX of GATT 1994 (WTO 1998c: 279 ff. (para. 17, p. 283, and para. 62, p. 299)). The United States appealed to the WTO Appellate Body, invoking in particular Article XX(g) to justify the legality of its measures.

Article XX(g) permits, as an exception to the GATT rules, measures 'relating to the conservation of exhaustible natural resources if such measures are made effective in conjunction with restrictions on domestic production or consumption'. In appraising Section 609 under Article XX of GATT 1994 the Appellate Body followed a three-step analysis. First, the Appellate Body asked whether the Panel's approach to the interpretation of Article XX was appropriate; and it concluded that the Panel's reasoning was flawed and 'abhorrent to the principles of interpretation we are bound to apply' (WTO 1998c, paras. 112–24 at 121). Second, the Appellate Body asked whether Section 609 was 'provisionally justified' under Article XX(g). Invoking the concept of 'sustainable development', it found that it was so justified (WTO 1998c, paras. 125–45). And third, it asked whether Section 609 met the requirements of the introduction to Article XX, concluding that it did not because the US actions imposed an 'unjustifiable discrimination' and an 'arbitrary discrimination' against shrimp to be imported from India, Malaysia, Pakistan, and Thailand.

It is in relation to the second and third steps that the Appellate Body invokes the principle of 'sustainable development', as an aid to interpretation. The Appellate Body's approach is premised upon an application of the 'customary rules of interpretation of public international law' as required by Article 3 paragraph 2 of the Understanding on Rules and Procedures Governing the Settlement of Disputes. These rules 'call for an examination of the ordinary meaning of the words of a treaty, read in their context, and in the light of the object and purpose of the treaty involved' (WTO 1998b: para. 114). It is these customary rules which the Panel failed to apply, leading to the conclusion at step one that the Panel's approach was flawed. It is then in relation to step two that the Appellate Body initially invokes the principle of sustainable development, in determining whether the measures taken by the United States are 'provisionally justified'. As a 'threshold question' the Appellate Body has to decide whether Section 609 is a measure concerned with the conservation of 'exhaustible natural resources', in the face of the argument that the term refers only to finite resource such as minerals, and not biological or renewable resources such as sea turtles (which, it was argued, fail to be covered by Article XX(b)). This argument was rejected by the Appellate Body. It ruled that Article XX(g) of GATT 1994 extends to measures taken to conserve exhaustible natural resources, whether living or non-living, and that the sea turtles here involved 'constituted "exhaustible natural resources" for the purpose of Article XX(g) of the GATT 1994' (WTO 1998*b*: para. 134). In reaching that conclusion, it stated that Article XX(g) must be read by a treaty interpreter 'in the light of contemporary concerns over the community of nations about the protection and conservation of the environment' (WTO 1998*b*, para. 129).

Referring to the Preamble to the 1994 WTO Agreement, the Appellate Body noted that its signatories were 'fully aware of the importance and legitimacy of environmental protection as a goal of national and international policy' and that the Preamble 'explicitly acknowledges "the objective of sustainable development"' (WTO 1998*b*, para. 129). This, says the Appellate Body, is a 'concept' which 'has been generally accepted as integrating economic and social development and environmental protection' (WTO 1998*b*, para. 129 with n. 107). According to the Appellate Body this conclusion is supported by numerous modern international conventions and declarations, including the UN Convention on the Law of the Sea. It follows that the sea turtles at issue were an 'exhaustible natural resource' and they were highly migratory animals, passing in and out of the waters subject to the rights of jurisdiction of various coastal states on the high seas. The Appellate Body then observes

Of course, it is not claimed that all populations of these species migrate to, or traverse, at one time or another, waters subject to United States jurisdiction. Neither the appellant nor any of the appellees claims any rights of exclusive ownership over the sea turtles, at least not while they are swimming freely in their natural habitat— the oceans. We do not pass upon the question of whether there is an implied jurisdictional limitation in Article XX(g), and if so, the nature or extent of that limitation. *We note only that in the specific circumstances of the case before us, there is a sufficient nexus between the migratory and endangered marine populations involved and the United States for the purpose of Article XX(g).* (WTO 1998*b*, para. 133, emphasis added)

The concept of 'sustainable development' is not expressly invoked to justify this potentially far-reaching conclusion as to the nexus between the sea turtles and the United States. Nevertheless, the concept appears to inform that conclusion, apparently establishing the necessary link between the interest of the United States in the proper conservation of a distant natural resource located from time to time outside its jurisdiction, and the finding that Section 609 is 'provisionally justified' under Article XX(g). Although the Appellate Body claims that it does 'not pass upon the question of whether there is an implied jurisdictional limitation in Article XX(g)', its conclusion appears hardly consistent with such a limitation. Between the lines, then, the concept of 'sustainable development' (and the need to integrate economic and social development and environmental protection) appears to have been implicitly invoked to extend (by interpretation) the jurisdictional scope of Article XX(g). If this is correct, then 'sustainable development' has a significant substantive element. This marks a significant move away from the approach of the earlier *Tuna Dolphin* panels (Sands 1995) and an opening which could, depending on your perspective, either strengthen global environmental objectives or contribute to unwarranted interferences by one state in the affairs of another. Having found that the US measures are 'provisionally justified', the Appellate Body then moves on to the third step of its analysis, namely whether Section 609 is consistent with the requirements of the introduction to Article XX. In my view the Appellate Body rightly concludes they are not, because the measures are applied in an unjustifiable and arbitrarily discriminatory manner. Of interest, however, is the fact that the Appellate Body invokes 'sustainable development' again, this time in the context of its conclusion that Section 609 is an 'unjustifiable' discrimination (WTO 1998*b*, paras. 161 ff. at 176). In the introduction to this part of its analysis, the Appellate Body revisits the Preamble to the WTO Agreement, noting that it demonstrates that WTO negotiators recognized 'that optimal use of the world's resources should be made in accordance with the objective of sustainable development', and that the preambular language, including the reference to sustainable development 'must add colour, texture and shading to our interpretation of the agreements annexed to the WTO Agreement, in this case the GATT 1994. We have already observed that Article XX(g) of the GATT 1994 is appropriately read with the perspective embodied in the above preamble' (WTO 1998*b*, para. 153). In support of the relevance of 'sustainable development' to the process of interpretation of the WTO Agreements, the Appellate Body then invokes the 1994 Decision of Ministers at Marrakesh to establish a Permanent Committee on Trade and Environment. That Decision refers, in part, to the consideration that 'there should not be . . . any policy contradiction between . . . an open, non-discriminatory and equitable multilateral trading system on the one hand, and acting for the protection of the environment, and the promotion of sustainable development on the other' (WTO 1998*b*, para. 154). The Appellate Body notes that the terms of reference for the establishment by this Decision of the Committee on Trade and Environment, which makes further reference to the concept of sustainable development, specifically refers to Principles 3 and 4 of the Rio Declaration on Environment and Development (WTO 1998*b*, para. 154).

This is all by way of introduction. There is no further reference to the concept of sustainable development, at least explicitly. Why then has it been invoked by the Appellate Body? No clear answer can be given to that question. However, it appears that 'sustainable development' informs the conclusion that the United States' measures constituted an unjustifiable discrimination: Section 609 established a rigid and unbending standard by which United States officials determined whether or not countries would be certified, and while it might be quite acceptable for a government to adopt a single standard applicable to all its citizens throughout that country, it was not acceptable, in international trade relations, 'for one WTO member to use an economic embargo to require other Members to adopt essentially the same comprehensive regulatory programme, to achieve a certain policy goal, as that in force within that Member's territory, without taking into consideration different conditions which may occur in the territories of those other Members' (WTO 1998*b*, para. 164). Shrimp caught using identical methods to those employed in the United States had been excluded from the US market solely because they had been caught in waters of countries that had not been certified by the United States, and the resulting situation was 'difficult to reconcile with the declared [and provisionally justified] policy objective of protecting and conserving sea turtles' (WTO 1998*b*, para. 165). This suggested that the United States was more concerned with effectively influencing WTO members to adopt essentially the same comprehensive regulatory regime as that applied by the United States to its domestic shrimp trawlers. Moreover, the United States had not engaged the appellees 'in serious, across-the-board negotiations with the objective of concluding bilateral or multilateral agreements for the protection and conservation of sea turtles, before enforcing the import prohibition' (WTO 1998*b*, para. 166).

The failure to have a priori consistent recourse to diplomacy as an instrument of environmental protection policy produced 'discriminatory impacts on countries exporting shrimp to the United States with which no international agreements [were] reached or even seriously attempted' (WTO 1998*b*, para. 167). The fact that the United States negotiated seriously with some but not other members that exported shrimp to the United States had an effect which was 'plainly discriminatory and unjustifiable'. Further, different treatment of different countries' certification was observable in the differences in the levels of efforts made by the United States in transferring the required TED technology to specific countries (WTO 1998*b*, para. 175). Moreover, the protection and conservation of

highly migratory species of sea turtles demanded 'concerted and cooperative efforts on the part of the many countries whose waters [were] tra-
versed in the course of recurrent turtle migrations' (WTO 1998*b*, para.
168). Such 'concerted and cooperative efforts' were required by *inter alia*
the Rio Declaration (Principle 12), Agenda 21 (para. 2.22 lit.(i)), the
Convention on Biological Diversity (Article 5), and the Convention on the
Conservation of Migratory Species of Wild Animals. Further, the 1996
Inter-American Convention for the Protection and Conservation of Sea
Turtles provided a 'convincing demonstration' that alternative action was
reasonably open to the United States, other than the unilateral and
non-consensual procedures established by Section 609 (WTO 1998*b*, para.
171).

   And finally, while the United States was a party to the 1973 Convention
on International Trade in Endangered Species of Wild Fauna and Flora
(CITES), it had not attempted to raise the issue of sea turtle mortality in
relevant CITES committees, and had not signed the Convention on the
Conservation of Migratory Species of Wild Animals or the 1982 United
Nations Convention on the Law of the Sea or ratified the 1992 Convention
on Biological Diversity (WTO 1998*b*, para. 171 with n. 174). The concept
of 'sustainable development' appears to have been invoked to provide
'colour, texture and shading' to the concept of an 'unjustifiable discrimination' in the introduction to Article XX. That in turn allows the Appellate
Body to reach out to these other, non-trade instruments to ascertain what
are the minimum standards to be met before discriminatory measures such
as those to be found in Section 609 may be justified under Article XX. In
this way 'sustainable development' has—beyond its substantive use in
relation to the meaning of Article XX(g)—a procedural element, namely
the requirement that appropriate diplomatic means, including those available within relevant multilateral agreements, be exhausted before unilateral measures may be taken.

CONCLUSIONS

To what extent is a reader of these two decisions enlightened about sustainable development? And why should we care what these two bodies
have to say?

   Both the ICJ and the Appellate Body refer to sustainable development as
a 'concept'. Both treat it as having a status in international law, in the sense
that it is invoked as part of a legal analysis to justify a legal conclusion.
Neither body explores its international legal status, whether as customary
or as conventional law. As Professor Lowe (1999: 36) has written in relation

to the ICJ's approach, the application of the concept of sustainable development has not been dependent 'upon proof of *opinio juris*, of the generality of state practice, or of compliance in any other way with the traditional requirements for the creation of customary international law'.

Neither body adds significantly to our sense of what it is or what role it has in the international legal order, beyond indicating that in normative terms it may have both procedural and substantive consequences. Yet both bodies apparently use 'sustainable development' as a significant aid to assist in reaching fairly radical conclusions. For the ICJ, 'sustainable development' contributes to the construction of a juridical bridge across which the majority on the bench marches to justify its conclusion that although the 1977 Treaty between Hungary and (Czecho)Slovakia requires the construction of two barrages and remains in force today, it does not now require Hungary to participate in the building of a second barrage. In effect, 'sustainable development' is utilized by the Court to rewrite the 1977 Treaty, to justify an interpretative conclusion which would not on its face be the outcome of its earlier prior analysis, or the traditional application of the law of treaties. The emergence of 'sustainable development' is a post-Rio fundamental change of circumstances which was not present in 1989 so as to justify Hungary's original suspension of works.

For the WTO Appellate Body, 'sustainable development' provides the 'colour, texture and shading' to permit interpretation of the GATT 1994 text which legitimately permits one state to take measures to conserve living resources which are threatened by actions in another state, subject to a need to exhaust multilateral diplomatic routes which may be available. This too is a far-reaching conclusion which breaks with prior international practice and for which little, if any, international precedent can be found.

From these two cases it appears, then, that 'sustainable development' remains as elusive as ever, generally requiring different streams of international law to be treated in an integrated manner. In the words of ICJ Judge Weeramantry (ICJ Reports, 1997), it aims at harmonization and reconciliation with a view to avoiding 'a state of normative anarchy'. The jurisprudence of those two bodies has not added greatly to our understanding of 'sustainable development': we do not know with a great deal more certainty what it is, or what international legal status it has, or in what precise way it is to be made operational, or what consequences might flow from its application. What we do know is that two important international judicial bodies have been prepared to invoke it to justify or support conclusions with consequences which challenge some basic tenets of traditional international law and are potentially far-reaching. At the very least, these two cases indicate that the 'concept' or 'principle' of 'sustainable development' has gained legal currency and that its consequences will be felt more rather than less widely. One can therefore expect 'sustainable development' to be relied upon in other fora, perhaps to justify the integration of environmental considerations into foreign investment protection agreements (for example, in the context of foreign investment disputes) or the integration of developmental considerations into the application of human rights norms.

Why should we care what these two bodies have done? For a start, their approach may well be picked up and relied upon elsewhere. There is already some indication that this is happening, for example in relation to the rights of civil society to participate in various international processes. This aspect of sustainable development, reflected in Principle 10 of the Rio Declaration, has been picked up by the WTO Appellate Body, even before the *Shrimp Turtle* case. The Appellate Body has expanded rights of *amicus* intervention to non-governmental organizations and corporations (WTO 2000, at para. 42): 'We are of the opinion that we have the legal authority under the DSU [Understanding on Rules and Procedures Governing the Settlement of Disputes] to accept and consider *amicus curiae* briefs in an appeal in which we find it pertinent and useful to do so. In this appeal, we have not found it necessary to take the two *amicus curiae* briefs filed into account in rendering our decision.' The Appellate Body's approach has now been picked up in a recent decision of an international arbitral tribunal established under the auspices of the dispute settlement system of the North American Free Trade Agreement (NAFTA), recognizing in principle the right of non-governmental organizations to file *amicus* briefs in such proceedings (*Methanex Corporation v United States of America*, 15 January 2001, at para. 53).

I would not wish to fall into the lawyers' common trap of overemphasizing the law-creating significance of these decisions, or indeed of their overall impact on human behaviour. But it is plain that in these two cases the international judiciary has had an impact which is likely to be felt more widely. These international courts are filling a gap left by states, which in the negotiation and legislative process are often unable to agree on definitive positions, and adopt ambiguous statements.

Into this void step the courts. Why is this relevant to governance? As Vaughan Lowe puts it, the ICJ 'could have managed without "the concept of sustainable development"; but it chose instead to refer to the concept and, by so doing, to open up the possibility of the development of the concept as a framework for the reconciliation of conflicts between development and environmental protection when they come before it'. He correctly observes that its reliance was premised upon the Court's perception that the concept had 'international acceptance', and that the content of sustainable development had 'international acceptance'. Lowe

(1999: 35) disagrees with the Court's approach, observing that the concept is not created by the traditional combination of state practice and *opinio juris*: it is, he says, 'essentially a judicial rule, created by judges and under their control'. The judges have given it a normative status, and the fact that they are willing to treat with it confirms that normative status.

This raises important issues about the role of the judiciary, issues which are much discussed in the United Kingdom following the incorporation into English law of the European Convention on Human Rights. Is it the function of judges to divine the will of Parliament (in the United Kingdom) or of states (in the international community)? The line between the application of the law and its development is a fuzzy one, especially in the field of environment and development. The approach raises questions about who the judges are (and how they are appointed), their independence, and their accountability. It raises questions about the best way to interpret ambiguous language: is it by reference to original meaning and intent, or by reading text in harmony with evolving standards of decency or morality? In relation to sustainable development, on either approach international judges will find themselves in difficulty and bringing to bear their own system of values.

The concept of sustainable development, whatever its legal status, whatever its normative value, is ambiguous and open-ended and is capable of being applied in opposing ways, as the arguments of Hungary and Slovakia demonstrated. There is not yet an international legislature or executive which can descend to the minutiae of determining what is and what is not consistent with sustainable development. There is no international equivalent to the English rule which says that the 'promotion of the use or proliferation of cesspools' is inconsistent with the attainment of sustainable development (see note 1). In this context the international judges will necessarily have to step in, analysing, interpreting, and applying. The emergence of the concept of sustainable development thus coincides with the emergence of an international judiciary, with all that implies for the governance of a sustainable world.

REFERENCES

Foundation for International Environmental Law and Development (1993). 'Report of the consultation on sustainable development: the challenge to international law'. *Review of European Community and International Environmental Law*, 2/4: 11–116.

ICJ (International Court of Justice) Reports (1996). 226–593.

ICJ Reports (1997). 7–241.

International Legal Materials (1999) 38: 118–75.

Lowe, V. (1999). 'Sustainable development and unsustainable arguments', in A. Boyle and D. Freestone (eds), *International Law and Sustainable Development.* Oxford: Oxford University Press.

Methanex Corporation v United States of America, Arbitration under Chapter 11 of NAFTA and the UNCITRAL Arbitration Rules, Decision of 15 January 2001.

R (On the Application of Anglian Water Services Ltd) v Environment Agency, Queen's Bench Division, Judgment of 24 October 2000, Tomlinson J.

Sands, P. (1994), 'International law in the field of sustainable development'. *British Yearbook of International Law*, 65: 303—81.

Sands, P. (1995). *Principles of International Environmental Law.* Manchester: Manchester University Press.

Sands, P., Mackenzie, R., and Shany, Y. (1999). *The Manual on International Courts and Tribunals.* London: Butterworths.

Stec, S., and Eckstein, G. (1997). 'Of solemn oaths and obligations: environmental impact of the ICJ's decision in the case concerning the Gabcíkovo–Nagymaros project'. *Yearbook of International Environmental Law*, 8: 41–50.

United Nations (1992). 'The Rio Declaration on Environment and Development'. *International Legal Materials*, 31: 874–80.

WTO (World Trade Organization) (1998a). Appellate Body Reports, AB–1998–4, 12 Oct.

WTO (1998b). *Report*, WT/DS58/AB/R. 12 Oct. http://www.wto.org/english/tratop_e/dispu_e/58abr.doc

WTO (1998c). *Report*, WT/DS58/R. 15 May. http://www.wto.org/english/tratop_e/dispu_e/58r01.pdf

WTO (2000). United States—Imposition of Countervailing Duties on Certain Hot-Rolled Lead and Bismuth Carbon Steel Products Originating in the United Kingdom, AB–2000–1, WT/DS138/AB/R. 10 May.

7

The Climate Change Regime: Can a Divided World Unite?

Joyeeta Gupta

INTRODUCTION: EVIDENCE OF A DIVIDED WORLD

THE end of the second millennium marks the beginning of the trend towards globalization on the one hand and, possibly, marginalization and disempowerment on the other hand. Despite the hype of globalization and the illusion of one world with shared hopes and concerns, there is increasing fear of marginalization of the South via depressed commodity prices, export barriers for Third World commodities, high international interest rates (South Centre 1993), foreign debt (George 1992), disempowerment of those not connected to the World Wide Web, and monopoly positions held by companies belonging to rich countries (Amin 1993). The wealth of the South has been in its natural resources and cheap labour, which were seen as attractive to colonial masters. In the post-colonial period the South sees the North as trying to claim authority over these resources through conventions on endangered species (see Bajaj 1996), biodiversity (see Lappe and Bailey 1999; Serageldin 1999; Shiva 1993), agreements on forestry (Gadgil and Guha 1997; Leow 1997), or trying to find ways of dumping toxic wastes on them (Third World Network 1989; Bhutani 1996) or old technologies. They see competition from their cheap products being countered through import tariffs (e.g. on textiles in the European Union) or social and eco-labelling schemes.

This article has been written on the basis of research being carried out in the context of Climate Change and the Law of Sustainable Development, a project of the Vrije Universiteit Amsterdam.

Against this background, this chapter explores the nature of international relations on global environmental problems such as climate change and tries to expose the growing divide between the 'ideal' of a global community, and the 'reality' of the fractures in the global community. It discusses how power inequalities exacerbate rather than minimize the fractures; and examines the potential of socializing behaviour in equalizing and/or perpetuating such inequalities through the medium of international law. It discusses global environmental issues in the context of North–South relations, because this is, in my view, the most serious challenge to the global community today (cf. Roberts and Kingsbury 1993).

Before going further, it may be useful to understand what 'the North' and 'the South' actually refer to. The North refers to the rich and relatively developed countries (including some former Eastern bloc countries) and the South consists of the rest, mostly poor and powerless, and the gap between the two is growing (UNDP 1996; UNEP 2000). The differences are not just of wealth, but also of historical circumstances. It would not, however, be out of place to mention that the North–South division is not as clear as their negotiating positions on climate change and other issues. There are many countries that ostensibly wish to belong to the club of rich countries, but their economic status and governance patterns have more in common with the developing countries. This is especially true of some of the former Eastern bloc countries. At the same time there are countries associated with the developing world that are as rich as the richest developed countries—notably Singapore. These memberships tend to blur the North–South negotiating patterns and reduce the credibility of the positions taken by the blocs. However, despite these aberrations there are clear distinctions between those in the rich world and the powerless in the developing world. This is not to deny that countries have issue-specific interests in different environmental regimes.

The most common criticism in literature on North–South issues is that there are vast differences within the North and within the South, and that focusing on North–South problems would only mean postponing the incremental measures that can be taken now to address global environmental problems. Further, some argue that the North–South perspective is actually damaging since it polarizes the issues and brings unnecessary issue linkages to debt, poverty, and technology transfer (Pearce and Perrings 1995). Echoes of this perspective are to be found in a range of legal, economic, and political science literature (Benedick 1993; Sebenius 1993). But I believe that such a perspective is flawed because there is no such thing as an objective problem (see Hisschemöller 1993) that needs to be depoliticized in order to be addressed. All problems are socially constructed and inherently political. Second, global environmental and economic problems are perceived as North–South problems by those actively engaged in the field, and especially by Southern intellectuals (Ramphal 1983; Nyerere 1983; South Commission 1990; Agarwal *et al.* 1992, 1999). Third, since solutions are often framed with a specific policy problem in mind, if these solutions do not match the diagnosis of the problem in other countries, the solutions are unlikely to be real 'solutions' that can be implemented. This means that we are in effect solving the wrong problem. Fourth, every time international relations scholars ignore the arguments being made by the South as inaccurate, irrelevant, and political, they are handicapping their own ability to search for solutions to global environmental problems. In doing so, they are unable to perform adequately the function of truly 'independent' scientists who are able to provide scientifically sound advice on how global problems can be solved.

While acknowledging that there are vast differences both in the North and the South, I believe that there is a serious need to examine, at least at the theoretical level, the implications of the divide for the future of global institutions and methods for addressing environmental issues. The key research question that flows from the above discussion is: Can the issue of global climate change be addressed without dealing with the relevant aspects of the North–South debate?

In addressing this question, this paper is based on more than 450 interviews primarily with stakeholders and negotiators from the developing world over the last eight years; and a literature survey not only of articles published in scientific journals, but also of popular journals, newspapers, government reports, and statements. The research is also based on firsthand observations of the way in which negotiators negotiate during the climate change negotiations.

CLIMATE CHANGE: A SHORT RECAPITULATION OF THE NEGOTIATIONS

The climate change problem refers to the greenhouse gases (like carbon dioxide, nitrous oxide, and methane) emitted during the burning of fossil fuels worldwide, among other activities. These gases have the potential to absorb some of the long-wave solar radiation re-emitted from the surface of the earth. The increased accumulation of these gases in the atmosphere is projected to lead to an enhanced warming of the earth's atmosphere. This enhanced greenhouse effect is expected to lead to rising sea levels, melting glaciers, rising temperatures, and, *inter alia*, extreme weather events.

There is no real dispute regarding the increase in the level of the emissions of greenhouse gases. The atmospheric concentration of carbon dioxide and methane has increased by 31 per cent and 150 per cent respectively since 1750. The surface temperature on the earth has increased by 0.6 ± 0.2 degrees centigrade in the last century; and this increase is likely to be the highest experienced in the last thousand years (IPCC 2001). There is also evidence that the sea level has increased and that the precipitation levels and patterns have changed. The Intergovernmental Panel on Climate Change (an international body of scientists and policymakers) thus indicates that there is a significant problem that needs to be addressed. While there is no dispute regarding these facts, there is some dispute regarding the cause–effect nature and the prediction of future impacts, since the climatic system is a non-linear system and there are many feedback effects, some of which are poorly understood. In particular, the issue of sinks raises significant questions. Sinks have the capacity to absorb greenhouse gases, but they may also release such gases. Thus forests absorb carbon dioxide, but deforestation also leads to a release of carbon dioxide. The precise nature of sinks (forests, soils, oceans, etc.) and how they function is less well understood than the nature of industrial sources of emissions, which can be relatively easily calculated.

The climate change issue was first signalled by the developed countries and the United Nations Framework Convention on Climate Change (FCCC 1992) was adopted in 1992. The Convention called on the developed countries to stabilize their greenhouse gas emissions at 1990 levels by 2000 and to provide technologies and financial assistance to developing countries to help them adapt to the potential impacts of climate change and to try and reduce the rate of growth of their emissions. The Convention was a major step forward. However, although there were five sets of principles adopted to ensure that countries would take cost-effective action in accordance with their common but differentiated responsibilities and respective capabilities, the precise degree of importance to be given to the principles and how they were to be interpreted was left open (Bodansky 1993; Gupta 1997). The reluctance of the United States to adopt a legally binding stabilization target led to a weakening of the legal wording of the text into a plea to the developed countries to reduce their emissions (Sands 1992). Although the developing countries had argued in favour of an independent financial mechanism to generate resources for them, the developed countries only agreed to provide financial resources if such a mechanism would be located at the Global Environment Facility of the World Bank, United Nations Environment Programme, and the United Nations Development Programme (Gupta 1995). At the same time there were no clear quantitative commitments regarding the nature of the assistance to be provided to the developing countries. But the rhetoric of leadership was included (Gupta 1998).

The Convention was ultimately ratified by more than 180 countries and entered into force in 1994. However, addressing the climate change problem is not quite as simple as it sounds. Reducing greenhouse gas emissions in effect implies reducing energy production from fossil fuels and restructuring the energy production and transport systems in countries. This is associated in the minds of many with considerable economic loss, especially for the vested energy interests, but also for the economy as a whole. In the period following 1992, pessimism set in and most of the developed countries found themselves experiencing relatively high economic growth and high greenhouse gas emissions. Vested interests blocked measures to restructure economies and by 1996 it was increasingly clear that the appeal made by the Convention to the developed countries to bring back their emission levels to 1990 levels by 2000 was unlikely to succeed. In the meanwhile negotiations on the follow-up to the climate convention were under way and the global community decided in 1995 that legally binding quantitative commitments were needed to encourage action. In the absence of legally binding commitments few countries were willing unilaterally to restructure their energy systems because of the costs that it would entail and especially because of fear that this would affect their ability to compete internationally. In the meanwhile 'aid fatigue' was setting in; very little by way of technology transfer was undertaken, and additional resources for the financial mechanism were very limited. In an effort to stimulate some cost-effective emission reduction the instrument of so-called Joint Implementation was promoted. The term was introduced in the Convention but not defined. For those who were pushing for the inclusion of this term, it implied that a developed country could invest in a developing country in return for emission reduction credits so generated (Arts *et al.* 1994). Developing countries initially saw this mechanism as a way for the North to continue living its current lifestyle and to buy all the cheap emission reduction options in the South (Maya and Gupta 1996). Nevertheless, in the hard negotiations that followed, developing countries eventually agreed in 1995 to a voluntary pilot phase on Activities Implemented Jointly, a new name for what, in effect, would lead to Joint Implementation. During the pilot phase, countries could participate on a voluntary basis and no credits would be given to countries. The idea was that lessons learned from this pilot phase would be incorporated later in a full-fledged Joint Implementation scheme.

Much to the surprise of international observers, the third Conference of the Parties to the Climate Convention did adopt a protocol in Kyoto in 1997 (for details of the history, see Grubb *et al.* 1999; Oberthür and Ott 1999). This Protocol urged the developed countries jointly to reduce their emissions of six greenhouse gases by 5.2 per cent by the budget period 2008/12. The forty developed countries were assigned differentiated emission allowances or targets. The Protocol also allowed countries to implement emission reduction projects where it was cheapest to do so. Thus countries were allowed to invest in other countries in return for emission reduction credits. Such projects in developing countries and in other developed countries (mostly east and central Europe) were allowed via the so-called Clean Development Mechanism and the Joint Implementation Mechanism respectively. These mechanisms were introduced to reduce the burden on governments and to allow the private sector to share responsibility in reducing greenhouse gas emissions. The Protocol also introduced the potential to trade emissions among those countries with quantitative emission-related commitments. Thus a country underusing its emission allowances could sell to another country whose emission levels were in excess of its allowance.

This agreement was a considerable step forward from the Convention because it included binding quantitative commitments for the developed countries and mechanisms to support implementation (cf. Grubb *et al.* 1999; Oberthür and Ott 1999). However, it was also criticized because although the science seemed to suggest that substantial reductions in emissions were necessary, at Kyoto only a −5.2 per cent target by the budget period 2008/12 was agreed to. This target was, in itself, considered very weak because it included three new gases, gave countries the right to choose between alternative base years for the new gases, and allowed countries to account for sinks in their calculations of emissions, and because it could be achieved through the use of several flexibility mechanisms. These mechanisms included the Clean Development Mechanism, Joint Implementation, and emission trading, explained above. It was feared by developing countries and several environmental organizations that the use of such mechanisms would allow the actual emissions of the developed countries to increase, while those of the developing countries would also increase (although less than they would otherwise have done), leading to a total increase in emissions. At the same time, the individual targets of the developed countries also raised questions. In the run-up to the Kyoto negotiations the European Union had agreed that it would allow the poorer member states the right to increase their emissions and this increase would be compensated by high emission reductions by the richer member states. The rest of the developed world were incensed by this differentiation and used it as an excuse to justify their own commitments. As a result Norway, Australia, and Iceland were able to negotiate commitments that allowed them to increase their emission levels by 1, 8, and 10 per cent respectively. This was seen by the developing countries as a betrayal of the principle of leadership. Ukraine and Russia were able to negotiate emission targets higher than their current and possibly future emission levels; this implies that they can sell their unused emission allowances to other developed countries and is referred to in the literature as 'hot air'.

In the meanwhile the US Senate linked ratification of the Kyoto Protocol to 'meaningful participation' by the developing countries, although what such participation implied was left unclear (Byrd-Hagel Resolution 1998; Clinton 1997). This led to increased pressure on the developing countries to take on voluntary commitments. While the developing countries were able to avoid discussing voluntary (quantitative) commitments at Kyoto in 1997 and Buenos Aires in 1998, the increasing pressure angered most countries, especially in the light of the fact that the bulk of the developed countries had not managed to stabilize their own emissions of greenhouse gases by 2000 (Gupta 1998). However, Argentina and Kazakhstan agreed in principle to take on voluntary commitments on their own terms.

The precise nature of the flexible mechanisms was, however, unclear and most countries postponed ratification until there was more clarity and until the United States ratified, arguing that, without US ratification, it did not make economic or environmental sense for the rest of the developed countries to take action (Gupta and Grubb 2000). However, when President George W. Bush took over, he announced that the United States was going to withdraw from active participation in the regime. This was enough to unite the rest of the world in the adoption of a consensus document in Bonn in 2001, and the political expectation is that these countries will ratify the Protocol even without the United States. Whether this will, in fact, happen remains to be seen.

The Bonn Agreement is a political statement that establishes a special climate change fund, a fund for the least-developed countries, and a Kyoto Protocol Adaptation Fund to help developing countries to access technologies to reduce their emissions and adapt to climate change. It also defined how 'sinks' can be used in calculating emission levels. It adopted some rules for the flexibility mechanisms. The inclusion of a wider definition of sinks at the Marrakesh meeting in November 2001 and increased flexibility in implementing the mechanisms are other very sore points for the developing countries, and also for environmentalists, although clearly their agenda is different.

NORTH–SOUTH FRICTION IN RELATION TO CLIMATE CHANGE

Against the brief background on the development of the climate change regime, this section postulates that there are several dimensions to the gap between the North and the South, especially in relation to esoteric issues like climate change.

Frustration Regarding the Right to Development

Since the 1960s the developing countries have put the issue of the right to develop on the international agenda, arguing that, in order for this right to have any effect, many changes in the international economic and political system would have to take place. Their demand for a new international economic order was never met, and the law of development remains an elusive element within the growing field of the law of sustainable development (Schrijver 1995, 2001). With the new issues of climate change, biodiversity, genetically modified organisms, etc., the old argument of neo-colonialism rears its head again. With each new international development, there is the fear that the North will prevent the development of the South because it needs a market but not a competitor. At the same time the two competitive advantages—natural resources and cheap labour—of the South are being lost through the technological revolution (De Rivero 2001: 186). It is against this background that the following section should be read.

Greenhouse gas emissions are to a large extent linked to economic growth under current circumstances. The bulk of the emissions are of carbon dioxide and these are mostly emitted while burning fossil fuels. Thus fossil-fuel-based energy production and consumption are closely related to the emissions of greenhouse gases. Reducing such emissions is thus associated with a (temporary) fall in national income, and most countries are clearly unwilling to take unilateral action. The problem for the developing countries is that their economies are presently energy-starved. For the developing countries, with the hindsight not available to the West in the past, large-scale hydro-power and nuclear power is no longer the easy option it was for the West. The same is now true for fossil fuel power. This leaves only the very expensive solar and wind energy options. Most developing countries are plagued by the nagging doubt that such measures may cripple their economies and leave them in a new spiral of debt to global multinationals or may result in mortgaging the future of their countries through the sale of credits to the developed countries.

In the early days of the negotiations it was understood that since the developing countries were, by definition, poor and vulnerable, it could not be expected that they would reduce their emissions. Instead this poverty would increase their vulnerability to the potential impacts of climate change. Since it was the developed countries that had emitted the most greenhouse gases thus far, it was their responsibility to reduce their own emissions, and also to make room for some further economic growth in the developing countries. At the same time the developing countries were to be assisted with modern technologies and aid so that they could avoid taking the same greenhouse-gas-intensive development path taken by the developed countries. The substance of this argument was reflected in several political declarations (e.g. the Noordwijk Declaration 1989 and the SWCC (Second World Climate Conference) Declaration 1990 and in the United Nations Framework Convention on Climate Change. However, the rhetoric was not translated into legally binding quantitative, and hence measurable, commitments as explained above, and implementation was left to the goodwill of countries.

There are several reasons why the interim negotiations are seen as unfair by the developing countries. First, the North is seen as backtracking on its original promise to lead by reducing its own emissions since few have stabilized their emissions and/or ratified the Kyoto Protocol (Gupta 1998). Second, since the regime now allows for emission trading, the allocation of emission allowances takes on a new significance. Countries with emission allowances can sell these to others and thus encash their unused emission allowances. Thus allocating emission allowances to countries becomes a much more politically charged issue. In theory such allocations can take place on the basis of criteria such as per capita rights or per hectare rights, or on the basis of current pollution levels. In the Kyoto Protocol no criteria for allocation have been adopted and allowance levels were based on what countries were willing to accept, and these allowance levels are quite close to their existing pollution levels. Hence, the approach to dividing up the property rights in the Kyoto Protocol is seen to benefit the largest polluters since they have the largest allowances. In effect if the polluter takes action to reduce his pollution, the *polluter gets paid*, as opposed to the *polluter pays* (see Agarwal *et al.* 1999). Developing countries have no allowances and, hence, no restrictions as yet. On the one hand, this means they cannot sell their unused allowances, and, on the other hand, they are accused by the new Bush administration of not taking on commitments. If, instead, the emission allowances had been on the basis of a per capita principle, for example, this would imply that the largest polluters would have to pay the smallest polluters (generally poor countries) in order to buy emission allowances. This is seen by developing-country negotiators as consistent with human rights principles (each person should have the right to live in dignity), as consistent with the polluter pays principle (the polluter is to be held responsible for the problem and should pay for the damage caused), and as consistent with the ability to pay principle (the rich are held to be better able to bear the burden of taking action for the collective good). Such transfers of resources would in principle, if used wisely, provide developing countries with the resources to both adapt and invest in clean technologies. However, this would entail a high economic burden for the developed countries, besides disturbing the status quo between countries, and is seen as unacceptable (see Cooper 1998; Grubb *et al.* 1999). Since discussing principles that could underlie the allocation of emission allowances or commitments to all countries is a sensitive issue, negotiations tend to bypass the discussion of such principles.

Third, the issue of how countries should develop sustainably is not being discussed in any of the key fora. Although there is a focus on sectoral solutions such as the promotion of energy efficiency, the climate negotiations do not focus on the complex issues of production and consumption patterns (Cutajar 1997; Chatterjee and Finger 1994: 173).

Fourth, this also implies that there is no method of preventing the dumping of old technologies on the developing countries, and there is no reason to suppose that this will not occur. Who is to screen these technologies to see if they are compatible with environmental protection and that they do not put the developing countries on an unsustainable development trajectory?

Fifth, in the process of negotiations the choice is being made to take the easy route and provide flexibility to keep the developed countries on board. Emission trading and the Clean Development Mechanism were introduced to keep the United States on board; targets to increase emissions were allowed to keep Australia, Norway, and Iceland on board; the new definition of sinks was introduced in Bonn in 2001 to keep Canada, Australia, and Japan on board. This accommodation process serves to keep the alliance united, but it also serves to alienate the developing countries, who perceive that the process is not developing along specific principles and criteria, creating the precedent that if a country has the power to bargain effectively it will get concessions, and, as a corollary, if a specific sector (energy) has the power to influence decision-making it can avoid accountability.

Sixth, although not much has been achieved in terms of technology transfer thus far (TERI 1998), it is expected that the financial mechanisms will stimulate a transfer of expensive modern technologies to the developing countries in the hope of receiving emission credits and new markets. Such technologies seen as too expensive for the West may create increased dependency and indeed debt for the developing countries (cf. Henikoff 1997: 51).

Seventh, even assuming that the private sector really wishes to reduce the emissions of greenhouse gases, there is no guarantee that the processes set in motion will lead to long-term effectiveness since poverty and poor governance will remain critical issues and may lead to further degradation of the resource base in the developing countries.

In fact my interviewees stated that, instead of the North questioning its own production and consumption patterns, it was trying to reduce emissions via assistance to the South, which ironically may put the developing countries on the same development trajectory. At the same time the North was seen as putting pressure on the South to take on voluntary quantitative commitments, thus restricting growth.

I would thus argue that the North–South divide in the climate change issue is intimately linked to (the right to) economic growth. The theory I would like to postulate here is that the South fears that it will always be the underdog in international relations and that its right to development will never be more than rhetorical. This fear is reinforced every time a new issue is signalled from the North and the South is dragged into the discussions. It is also reinforced every time the South signals an issue (e.g. desertification), and when the issue is given at best lip-service by the countries of the North. Regarding the issue of climate change, the fear is that the days of using domestically available fossil fuels and fossil fuel technology will somehow be numbered and that this will cramp the development potential of the developing countries. In many ways the South itself contributes to the processes and negotiation outcomes that reconfirm this, as will be shown below.

The Theory of the Hollow Mandate of Developing Countries

It would not be out of place to discuss some of the difficulties faced by negotiators from the developing countries in the international negotiations. First, interviews reveal that most developing-country negotiators and policymakers face *six ideological dilemmas* on how to modernize without Westernizing, to survive without squandering resources, to request financial help without mortgaging the future, to empower the private sector to solve public problems, to ask for equity from the North without ensuring equity domestically, and to fight for short-term economic gain in the negotiations without making long-term losses (Gupta 2000*a*). These dilemmas hamper the ability of negotiators to respond effectively during the lively negotiations, especially when domestic dialogue and decision-making does not precede the negotiation process. For example, countries may hate the idea of importing toxic wastes because of the neo-colonialist and neo-imperialist implications but at the same time be tempted by the financial rewards that may accompany such wastes.

Second, most developing-country negotiators are affected by the *structural imbalance in knowledge*, where the bulk of the scientific work on issues signalled by the North is done in the North and is based on Western assumptions, theories, and analytical techniques and is not subject to adequate critical peer review from the South (cf. Kandlikar and Sagar 1999; Agarwal and Narain 1991, 1992; Gupta 1997, 2001*a*). The fear is that the data will somehow be used to emphasize the role of the South in the process (methane emissions from rice fields, carbon dioxide emissions from deforestation) and thereby distract from the role of fossil fuel use in the North. Controversies regarding emission scenarios, methane emissions and sinks (Agarwal and Narain 1991), and the cost of human life (Pearce *et al.* 1995) are seen as the tip of the iceberg of controversial science uncontrolled by effective peer review by Southern researchers.

Third, preparation in developing countries for international negotiations on global environmental issues signalled by the North is affected by lack of public awareness and the combined lack of concern of environmental non-governmental organizations, industry, and politicians, which lead to an absence of any social debate on how the trade-offs between different goals should be achieved. This is partly because such issues are likely to be defined in terms that are alien to domestic concerns in the South, and possibly compete with domestic priorities. Because they are defined in alien terms they are unlikely to mobilize grass-roots organizations; and those organizations that are mobilized are likely to be seen as suspect because many receive their funding from foreign agencies. Such global environmental issues often expose a government's inability and vulnerability and are therefore unlikely to be exploited for political purposes except to highlight the North–South dimensions.

Fourth, preparation for international negotiations on modern environmental problems signalled by the North, including climate change, is hampered by the lack of material issue linkages (e.g. with energy, transport, water management) made by policymakers and negotiators from the South. Issue linkages made by the South tend to be historical (since information on colonialism and imperialism is more easily available and accessible) and rhetorical (since there is information regarding the negotiating positions in other international treaties but not so much regarding the specific dimensions of the issue under discussion) rather than material. On occasion material issue linkages are made by individual countries.

Fifth, interviews reveal that domestic preparation is hampered by the 'four Ds': delegation of responsibility to a relatively low level; diminution of responsibility to discussion of North–South issues and not necessarily domestic energy and economic issues; downsizing the delegation to one or two delegates; and discontinuity in the delegation. To the extent that countries have a domestic policy-making process for international climate change negotiations, this process tends to be fractured and frequently more a 'matter of form than of content and strategy', as an interviewee put it. As the items on the international agenda get more and more complicated, domestic preparation glosses over the details and leads to an overall position which tends to be rhetorical.

Sixth, preparation for international negotiations on modern environmental problems signalled by the North, including climate change, is hampered by the qualitative, elitist, and diplomatic definition of national interests in the South. This implies that the negotiators tend to prepare a position that has mostly qualitative elements (need for technology transfer and finance), elitist definitions (mostly mirroring the ambitions of the elite: the desire to become like the West), and diplomatic wording. As one former ambassador put it: 'But on ecological issues, we take an ideological position. We don't like it, but it is a matter of negotiation.'

These are the six reasons why delegates from developing countries generally have a hollow negotiating mandate (Gupta 1997, 2000*b*) in the climate change regime, which includes general instructions and a bottom line. The exceptions are the OPEC group, which has a very well-defined agenda and negotiating strategy based on protecting its position as oil producers and exporters (Oberthür and Ott 1999), and, on occasion, the large developing countries and the forty small island countries (see e.g. AOSIS Protocol 1994).

The Defensive Negotiating Strategy of the South

How do the above observations influence the ability of the developing countries to negotiate? The hollow mandate inevitably leads to the adoption of a defensive negotiating strategy. Interviews, observations, and reviews in the literature reveal that such a strategy consists of a number of elements. One key element is ad libbing: negotiators make up the text as the negotiations proceed. They in general avoid making new proposals because these need to be based on detailed information and knowledge, tested and approved at home before they can be launched in public. Instead they stick to opposing proposals from the North on the basis that the North will only propose something that is in its own best interest and that history has shown that the North is likely to use the international order to prevent the development of the South. The negotiators also tend to vacillate between one ideological position and another, especially when side payments are offered. Even though the force of the arguments for keeping issues separate is apparent to them, they continue to see issues holistically within the context of North–South relations in general. After the negotiation is completed, the Southern negotiators feel cheated by the results because they feel that they were inadequately prepared, they had inadequate opportunity to influence the process, and they had no choice but to accept the outcome. The defensive strategy is not necessarily legitimate in the sense that it has broad-based political and possibly social support; but it is based on what may be referred to as proxy indicators of legitimacy, such as precedents borrowed from other issue areas and earlier meetings on the issue.

This defensive position is seen by developing-country negotiators as the best that can be done under the circumstances. It is seen as more legitimate than an impromptu position; it can be easily defended in the domestic context because it is consistent with other foreign policies; and it can, if used carefully, lead to concessions. On occasion the developing countries use a constructive strategy, but there are many perceived risks in such a strategy since it often lacks legitimacy; negotiators may lose their job for pursuing a risky course; and there is no guarantee that, if their proposals are accepted, they can persuade their own governments to implement them (Gupta 2000*b*).

The Theory of the Handicapped Coalition-Making Power of Developing Countries

Another critical problem facing developing countries is the difficulties experienced in coalition-building. There are 153 countries outside the developed world and one would expect that these countries could be quite a formidable group if they pooled their resources. But interviewees argue that they do not have much to pool—that many of them do not have good relations with their neighbours and that makes cooperation very difficult. They often have contradictory ideological positions, which also fluctuate from time to time. They suffer from a combined structural imbalance in knowledge. There is combined apathy and helplessness in the domestic situation. Negotiators tend to focus on North–South issues. In terms of process, they have limited staying power and resources to participate in intra-sessional preparations. The process of combining vaguely identified national interests leads to an even vaguer position at G–77 level. One diplomat put it like this: 'The South's approach may not be productive—but it's the only way to have a platform. It would be much better if the South negotiated on economic issues, but the capacity and ability to articulate these is much less.' Over and above these problems are the actual difficulties in accessing means of communication. Thus the common refusal to discuss a proposal on voluntary commitments for developing countries led to its removal from the final text of the Kyoto Protocol. I would argue that this adds up to a handicapped coalition-building power for the developing countries.

This handicapped coalition-building power leads to a brittle defensive strategy. The strategy is brittle because of the present lack of good leadership in the South and because many developing countries are confused about whether to negotiate in collaboration with the G–77 or whether to make individual deals with the North. Collaboration strengthens the negotiating power of the individual country, but individual deals can be attractive. Many fear that such deals are used to divide and rule the susceptible South through the use of the word 'voluntary'; through selective use of side payments and issue linkages; through selective use of 'punishments' in other issue areas; and because the countries that are likely to 'graduate' to developed-country status are afraid that they will also have to adopt such commitments in the future. The use of the word 'voluntary' is seen as an effective tool to divide the South, to 'make everyone cheat' and to help countries that are 'prepared to break rank' to get 'some crumbs'. One interviewee said, 'personally there is no problem in voluntarily signing up for action. I mean, who are you to tell others what they should do? But this leads to divide and rule.' Another point raised by negotiators is that very often they have to break ranks because of pressure from developed countries with which they have bilateral relations (Gupta 2000*b*).

The Theory of Handicapped Negotiating Power

I would further argue that the hollow mandate and the handicapped coalition power results in a handicapped negotiating power with a brittle (as opposed to a resilient) and threadbare (as opposed to solid) defensive strategy during the international negotiations (Gupta 2000*b*). The strategy of the developing countries is threadbare because they do not have the numbers, expertise, language skills, and staying power to negotiate effectively during all-day and -night sessions taking place simultaneously in two plenary halls, or in several so-called non-groups and groups, or in the corridors to compete with the developed countries. Nor do they have support in terms of being represented by well-informed high-level officials and politicians from their own countries, or their own scientists, non-governmental organizations, and industry with whom they can consult during the negotiations. The Kyoto Protocol was adopted long after several negotiators had already gone home on their non-refundable flights. The lack of adequate negotiators, and multiple negotiations on related issues taking place simultaneously in different rooms, does not provide for a neutral setting for such negotiations (cf. Schelling 1960: 31). As an interviewee put it, 'What happens inside the negotiation rooms determines very little. Late-night meetings in smoke-filled rooms, late calls home, corridor discussions, lunches, individual agreements outside the plenary are then sold to the plenary.'

The Theory of Structural Imbalance in Bargaining

When brittle, threadbare, and defensive strategies are used by the South (or other handicapped negotiators), this works well only when the other party is committed to addressing the problem. But if the other party is also using defensive strategies, the negotiations lead to rhetorical and symbolic statements, prevarication, and indeterminate language, but not problem-solving. When the developing countries use constructive strategies, this may imply additional costs and responsibilities for the North, and often leads them to adopt a defensive posture. In this position they may often resort to breaking down the proposal of the South into its component parts and then reconstructing the deal such that it no longer resembles the initial proposal of the South. For example, the Brazilian proposal for a Clean Development Fund to be created from fines levied on developed countries that were not in compliance metamorphosed into the Clean Development Mechanism, a new name for Joint Implementation. Thus constructive Southern proposals, when pitted against defensive Northern approaches, often lead to exclusion strategies when ideas are considered irrelevant for discussion (such as issue linkages to debt, poverty, etc.), or non-decisions (when issues are suppressed), or accommodation on paper (e.g. when vague commitments are made to provide technology transfer and funding) (cf. Bachrach and Baratz 1970). When the Northern countries use constructive strategies and the South is defensive, this leads to imposed decisions, decisionless decisions (when certain steps are taken leading in effect to the implementation of a decision, even though the decision was never really taken), and on occasion concessions for the South.

The negotiation outcomes are further aggravated when developed countries adopt incremental approaches (which can lead to decisionless decisions) and sectoral approaches (which can lead to non-decisions, which 'is a decision that results in suppression or thwarting of a latent or manifest challenge to the values or interests of the decisionmaker'; Bachrach and Baratz 1970: 44), and think globally (which can lead to irrelevant solutions at local level). As Southern interviewees explained, 'When we draft documents, we want the North to understand and not just hear. But it is very convenient for them to only hear and not understand.' 'So the Northern drafters use their legal language to misunderstand us and confuse us. So it becomes sometimes very difficult for us. We may say yes to the polluter pays principle or the sustainable development principle, only to discover later that it works against our interests and that they have defined it to suit their interests.'

The Competing Theories of Problem-Solving

Let us then return to the issue of problem-solving. Are the North and the South discussing the same problem? Hisschemöller's (1993) theory on structured and unstructured problems can be adapted to the international context. A structured problem is one in which there is consensus on the science and values of the problem. Conversely, there is no consensus on the science and values of an unstructured problem. In order to address unstructured problems, the problem needs to be structured. In the international context the North sees the climate change problem as merely a problem of sources of emissions and sinks; and sees the solution in reducing emissions and enhancing sinks. The problem is one of cost, and hence the problem needs to be addressed cost-effectively. On the other hand, the South sees the problem as one of lifestyles and production and consumption patterns, which can only be addressed by dealing with these lifestyles and their underlying ideologies. Since there is a problem of costs, they should be shared equitably, with those that caused the problem and with the ability to deal with the problem taking on the bulk of the responsibilities. From this perspective, the climate change problem is an unstructured problem that needs to be structured first as part of the process of problem-solving.

The North would, however, like to deal with it as if it were a structured problem by ignoring the perspective of the South. Then it would like to apply a realistic approach to problem-solving which involves breaking up complex issues into small manageable segments; to take the easy issues first; to try and come up with incremental solutions for the easy issues in order to keep the process from polarizing (Sands 1990; Greene 1996; Benedick 1993; Sebenius 1993). They argue that 'socially fragmented' policy and 'disjointed incrementalism' can be part of legitimate policy and that small improvements can lead to big changes in the long term (Braybrooke and Lindblom 1963).

Hisschemöller and Gupta (1999) argue instead that most of these complex North–South problems cannot be addressed adequately by pretending that the problems are structured. For 'structuring' the problems it becomes essential that there is a real dialogue between the parties, that controversial issues are discussed in an open sphere, and that norms and principles are developed that apply across the board to similar issues and similar situations. Such problems cannot be dealt with through linear, incremental solutions and by ignoring the systemic, integrated, and complex cause-and-effect relationships. The risk then is that the actual cause of the problem is never addressed (Gerlach 1992; Braybrooke and Lindblom 1963). In other words, while breaking problems down into small manageable segments creates the illusion that they can be addressed, the fact that these problems are possibly linked to poverty, debt, and the whole paraphernalia of issues that concern the South is more than likely to lead to further degradation of the resource base in the South (UNEP 2000, p. xxiii) and is not likely to make the South a willing partner in environmental treaty negotiations. Ramphal (1983) has argued that de-linking issues is not a strategy of collective self-reliance but instead the choice of the drop-out state that settles for a poor deal.

The Theory of Decreasing Legitimacy

International negotiations on modern complex environmental issues promoted by the developed countries tend to be threatened by the decreasing legitimacy of the international process and hence lack of incentive to comply (for details, see Gupta 2001*b*). International problem-solving in the area of environmental problems is undertaken generally through the adoption of declarations, and political and legal agreements within the framework of the United Nations, and using the rules and procedures of international law.

There are different visions of why international law works (Arend 1996; Beck 1996). Underlying all these visions are implicit and explicit assumptions about international law. These assumptions include the notion that the state is assumed to be the sole actor in international law (see Article 6 of the Law of Treaties), and that all states are sovereign and equal (see Article 2 of the UN Charter). There is an assumption that when states negotiate with each other, they are in effect talking about the same problem; that states send informed and well-prepared negotiators to the negotiations and cannot afterwards claim that they were not well informed (see Articles 47 and 48 of the Law of Treaties). There is also an assumption that the existence of clear rules of procedure will guarantee the rule of law and ensure that the negotiations are balanced; that the negotiation outcomes will be determinate and clear and legitimate and therefore have normative force (see also Franck 1990, 1995). There is also the assumption of *pacta sunt servanda*, that countries will negotiate in good faith and have the institutional capacity to implement the agreements adopted.

At the same time it is becoming increasingly clear that these assumptions are quite often wishful thinking. We live in plural societies, and social actors, epistemic communities, and civil society all want to influence the international policy-making process and do not believe that the state can represent their multiple views effectively. The notion of sovereign equality of states, while very useful legal fiction, is fiction, since countries differ vastly from each other, and providing each country with one vote does not imply that, for example, Kenya has as much say in international affairs as the United States. While there appears to be consensus, in that all countries are dealing with the problem of climate change, interviews reveal diverging definitions of the problem. As argued above, countries are often unable to send well-prepared negotiators to the negotiating table. Even though there are rules of procedure, the negotiations are far from balanced. The brittle, threadbare, defensive strategies of the South when pitted against the generally speaking realist, constructive strategies of the Northern countries lead to decisions that often favour the perspective of the developed countries, although in the process the developing countries may not themselves gain much. The outcome of the negotiations is often vague and the result of hard bargaining, and not all states have the institutional capacity to implement the agreements taken on by them. All this implies that when negotiations take place between states at different levels of development and on issues that are signalled by the North, the outcome of the negotiations may not have the desired degree of legitimacy and may thus have a lower compliance pull,[1] as the negotiators from the South may not feel as committed to the outcome.

The Theory of Regulatory Competition and First-Comers

In the previous section I argued that, on problems identified and defined by the developed countries, the developing countries are ill prepared to negotiate effectively. This section goes further to argue that in fact this situation is aggravated by the desire of specific developed countries to engage in regulatory competition (Héritier 1996; Héritier *et al.* 1996), i.e. to compete in promoting tools to address the problem that have already had success in their own countries. Individual developed countries try to globalize ('upload') domestic strategies because if they are successful in doing so, then the cost of domestic implementation for them is eventually much less, since they have experience and an institutional framework that is compatible with the international consensus. However, other countries, which have to implement these solutions, may find themselves facing expensive policies since they may have to develop the institutional infrastructure to do so.

Regulatory competition not only puts a strain on other developed countries that have to participate in the regime, but it increases the difficulties faced by developing countries in implementing such agreements. So, even if the developing countries have good substantive arguments, these are countered with pacification or accommodation strategies. The existence of

---

[1] The term 'compliance pull' refers to the capacity of a legal text to attract countries to comply with the contents. In international environmental law there is limited scope for forcing compliance—countries must themselves be motivated to comply with a legal commitment. The compliance pull is enhanced by the way the negotiations take place and by the wording of the negotiated text.

regulatory competition also increases the costs of implementation for the developing countries, so it further skews the equity in the regime. It also undermines the legitimacy of the eventual outcome because of the lack of domestic support and understanding for the negotiated agreement (Börzel and Gupta 2000).

Regulatory competition tends to benefit the first-comers in environmental issues. Castells (1999) argues, on the basis of her analysis of late-comers and first-comers in an environmental issue in Europe, that if the interests of the late-comers are not taken into account by the first-comers, then it will be very difficult to secure their commitment. First-comers are always able to benefit in international regimes since they can design the rules and these rules may also be in their own favour.

IMPLICATIONS FOR INTERNATIONAL LAW AND RELATIONS

The arguments above indicate that global relations appear to be irrevocably skewed against the interests of the South in situations where the issues on the negotiating table are signalled by the North, since the developing countries have a hollow negotiating mandate, a handicapped coalition-building power, and a handicapped negotiating power. The situation is further aggravated by the precedent set by earlier North–South relations, because of the decreasing legitimacy of the process as it gains in speed and complexity, and the tendency of the first-comers to engage in regulatory competition. This will continue to be the case as long as the issues are not defined in terms that can easily be prioritized in developing countries, as long as they are unable to benefit from modern communication techniques, and as long as they see themselves as continually running after the West. This will also continue to be the case as long as they do not have widespread domestic support and support from civil society in presenting their case. As long as the outcome of negotiations is resented by the negotiators from the South and has limited legitimacy, it will have a low compliance pull. This means that commitments flowing from such agreements will not be fully complied with by the developing countries.

Interviews reveal that this will lead to increased frustration in the South because it believes that the North will maintain the international status quo by exploiting the South's inability to negotiate and by using social pressure to gain the commitment of the South. The South believes that in the process poor precedents are being created, failed processes are being institutionalized, and the normative role of the law is being marginalized. Of course, it is not just the South that is becoming frustrated. The North, too, is getting tired of having to deal with the brittle, threadbare, defensive strategies of the South. It is becoming frustrated with the attempts of the South to link every issue with other past and present issues, is annoyed by the desire of the South to use every possible issue to change the status quo between countries, and feels that the poorer countries want to avoid doing anything to address the problem. The North is getting bored with the argument that all Southern ailments (corruption, poor governance, political stress, etc.) are caused by the North. Northern interviewees argue that in their view the South is taking a free ride and is unwilling to guarantee on what terms it will be willing to take action.

This increasing frustration on both sides often leads to polarization in the discussions, and if it carries on for another decade might lead to the erosion of faith in the long-term success of international regimes and treaties on modern environmental problems signalled by the North, in turn perhaps threatening the rule of law and civilized decision-making in the environmental field. Is this view too pessimistic? Developments in the climate change regime show that although the North was initially optimistic and willing to lead the way, the leadership paradigm became conditional. Since George W. Bush became President, the US position has hardened on the grounds that the Kyoto Protocol is unfair, too soft on developing countries, and too expensive for the United States. However, this has caused the rest of the world to unite. But in the wake of the recent terrorist incidents it appears that the US leadership has realized that it needs the United Nations and international law on its side to fight the terrorists. This may lead to a reversal in existing trends.

CONCLUSIONS: IS THERE POTENTIAL FOR UNITY?

The existence of a common enemy (climate change) is not in itself enough to unite the divided world. The problem has different scientific, political, and social dimensions for different countries. In the first part of this paper I asked the question: Can the issue of climate change be dealt with without dealing with the relevant North–South aspects? Clearly, if industry can identify a technology that can quickly make the world carbon-free, the issue of dividing the environmental space through the allocation of emission allowances between countries will become less urgent. However, the technological solutions are presently seen as expensive in the North and unaffordable in the South. Should these technological solutions be promoted by governments, the price will come down and make them more affordable for all. But there are political, financial, ideological, technological, and institutional challenges in pursuing such a path. These challenges lie in the perceived risks to the competitiveness of domestic industry if the developed countries were to restructure their economies by investing in modern greenhouse-friendly technologies. Moreover, such technologies are possessed by few developed countries, and this inevitably implies North–North transfers of resources to support the industries of competitive countries. Besides, countries are caught up in a technological lock-in, where existing lifestyles and infrastructures appear to be indispensable and countries and populations appear to be somewhat unwilling to change their habits. Nevertheless, technological and institutional transformation worldwide is necessary to deal with the problem.

The potential of a technological solution reduces the urgency of addressing the North–South issue in the context of the climate change debate. Yet, before such a solution can become a reality, the South will have to be able to engage effectively in the negotiations. However, as argued above, the South is at a major disadvantage in issues signalled by the North since it is always trailing behind in the negotiation process. For the South, this means that it needs to assess its own strengths and weaknesses. The recent global summit of the G–77 (2000) barely mentioned the environment. But the natural resources of the South need to be protected both to ensure food and water security domestically and also to enable the marketing of products internationally. This realization seems to have escaped the attention of the South. It also need to find ways to strengthen its coalition-building skills and to identify opportunities in international negotiations to push its case further. The North, too, needs to engage in dialogue with the South and to internationalize those same values that it holds so dear domestically. While there is much talk of political human rights, social and economic human rights are not supported adequately in the international terrain. While there is much talk of democratic and transparent policy-making practices domestically, inadequate emphasis is laid on democratic and transparent negotiation practices internationally. While there is much talk about justice and fairness domestically, equity is a stepchild in international relations. This calls for 'uploading' those very domestic principles that the West holds so dear onto the international arena. It calls for recognizing and using the issue links made by developing countries. It calls for levelling the international playing field.

In such situations, theorizing about how the design of the regime can be improved will result in practical failure unless such theories actually take into account the situations under which developing countries negotiate and try to make the system more transparent to these countries. This is especially necessary in the case of unstructured problems. This is not to deny that the diversity of the developing countries offers options for developing tailor-made solutions and coalitions without exacerbating the North–South divide. But that is not the issue under discussion in this chapter.

REFERENCES

Agarwal, A., and Narain, S. (1991). *Global Warming in an Unequal World: A Case of Environmental Colonialism.* New Delhi: Centre for Science and Environment.

Agarwal, A., and Narain, S. (1992). *Towards a Green World: Should Global Environmental Management be Built on Legal Conventions or Human Rights?* New Delhi: Centre for Science and Environment.

Agarwal, A., Carabias, J., Peng, M. K. K., Mascarenhas, A., *et al.* (1992). *For Earth's Sake: A Report from the Commission on Developing Countries and Global Change.* Ottawa: International Development Research Centre.

Agarwal, A., Narain, S., and Sharma, A. (1999). *Green Politics: Global Environmental Negotiations.* New Delhi: Centre for Science and Environment.

Amin, S. (1993). 'The challenge of globalisation', in South Centre (ed.), *Facing the Challenge: Responses to the Report of the South Commission.* London: Zed Books.

AOSIS Protocol (1994). *Draft Protocol Text,* A/AC./237/L.23, 27 Sept. Bonn: FCCC Secretariat.

Arend, Anthony C. (1996). 'Towards an understanding of international legal rules', in R. J. Beck, A. C. Arend, and R. D. Vander Lugt (eds), *International Rules: Approaches from International Law and International Relations.* Oxford: Oxford University Press.

Arts, K., Peters, P., Schrijver, N., and Sluijs, P. van (1994). 'Joint implementation from an international law perspective', in O. Kuik, N. Schrijver, and P. Peters (eds), *Legal and Economic Aspects of Joint Implementation to Curb Climate Change.* Dordrecht: Kluwer.

Bachrach, P., and Baratz, M. S. (1970). *Power and Poverty: Theory and Practice.* Oxford: Oxford University Press.

Bajaj, R. (1996). *CITES and the Wildlife Trade in India.* New Delhi: Centre for Environmental Law, WWF India.

Beck, R. J. (1996). 'International law and international relations: the prospects for interdisciplinary collaboration', in R. J. Beck, A. C. Arend, and R. D. Vander Lugt (eds), *International Rules: Approaches from International Law and International Relations.* Oxford: Oxford University Press.

Benedick, R. E. (1993). 'Perspectives of a negotiation practioner', in G. Sjostedt (ed.), *International Environment Negotiation.* Laxenberg: International Institute for Applied Systems Analysis.

Bhutani, S. (1996). *The Basel Convention and the Import of Hazardous Wastes to India.* New Delhi: Centre for Environmental Law, WWF India.

Bodansky, D. (1993). 'The United Nations Framework Convention on Climate Change: a commentary'. *Yale Journal of International Law,* 18: 451–588.

Borzel, T., and Gupta, J. (2000). 'A new North–South conflict? Regulatory competition in European and international environmental politics', Paper presented at the Conference of the European Concerted Action on the Effective Implementation of Environmental Law, Barcelona, 9–11 Nov.

Braybrooke, D., and Lindblom, C. E. (1963). *A Strategy of Decision: Policy Evaluation as a Social Process*. New York: Free Press.

Byrd-Hagel Resolution (1998). (Senate Resolution 98.) Congressional Record, 3 October 1997 (Senate), S10308–11. http://www.microtech.com.au/daly/hagel.htm

Castells, N. (1999). *International Environmental Agreements: Institutional Involvement in European Transboundary Air Pollution Policies*. Ispra: Joint Research Centre, European Commission.

Chatterjee, P. and Finger, M. (1994). *The Earth Brokers*. London: Routledge.

Clinton, W. J. (1997). Remarks by the President on Global Climate Change, Speech to the National Geographic Society, 22 Oct.

Cooper, R. N. (1998). 'Toward a real global warming treaty'. *Foreign Affairs*, 77 (2): 66–79.

Cutajar, M. Z. (1997). 'The road to Kyoto and an agreement that works'. *Climate Change Bulletin*, second quarter, 2.

De Rivero, O. (2001). *The Myth of Development*. London: Zed Books.

FCCC (1992). *United Nations Framework Convention on Climate Change*. Bonn: FCCC Secretariat.

Franck, T. M. (1990). *The Power of Legitimacy among Nations*. Oxford: Oxford University Press.

Franck, T. M. (1995). *Fairness in International Law and Institutions*. Oxford: Oxford University Press.

G–77 (2000). *Statement of the South Summit, Group of 77 Havana, 10–14 April 2000*. Havana: Group of 77.

Gadgil, M., and Guha, R. (1997). *This Fissured Land: An Ecological History of India*. Oxford: Oxford University Press.

George, S. (1992). *The Debt Boomerang: How Third World Debt Harms us All*. London: Pluto Books.

Gerlach, L. P. (1992). 'Problems and prospects of institutionalising ecological interdependence in a world of local independence', in G. Bryner (ed.), *Global Warming and the Challenge of International Cooperation: An Interdisciplinary Assessment*. Provo, Ut.: Brigham Young University.

Greene, O. (1996). 'Lessons from other international environmental agreements', in M. Paterson and M. Grubb (eds), *Sharing the Effort: Options for Differentiating Commitments on Climate Change*. London: Royal Institute of International Affairs.

Grubb, M., Vrolijk, C., and Brack, D. (1999). *The Kyoto Protocol*. London: Earthscan/RIIA.

Gupta, J. (1995). 'The global environment facility in its North–South context'. *Environmental Politics*, 4 (1): 19–43.

Gupta, J. (1997). *The Climate Change Convention and Developing Countries: From Conflict to Consensus?* Dordrecht: Kluwer.

Gupta J. (1998). 'Leadership in the climate regime: inspiring the commitment of developing countries in the post-Kyoto phase'. *Review of European Community and International Environmental Law*, 7 (2): 178–88.

Gupta, J. (2000*a*). 'Global environmental issues: impact on India', in S. N. Chary and V. Vyasulu (eds), *Environment Management: An Indian Perspective*. New Delhi: Tata McGraw Hill.

Gupta, J. (2000*b*). 'North–South aspects of the climate change issue: towards a negotiating theory and strategy for developing countries'. *International Journal of Sustainable Development* 3 (2): 115–35.

Gupta, J. (2001*a*). 'Effectiveness of air pollution treaties: the role of knowledge, power and participation', in M. Hisschemöller, R. Hoppe, W. Dunn, and J. Ravetz (eds), *Knowledge, Power and Participation*, Policy Studies Annual. Somerset, NJ: Transaction.

Gupta, J. (2001*b*). 'Legitimacy in the real world: a case study of the developing countries, non-governmental organisations and climate change', in J.-M. Coicaud and V. Heiskanen (eds), *The Legitimacy of International Organizations*. Tokyo: United Nations University Press.

Gupta, J., and Grubb, M. (eds), (2000). *Climate Change and European Leadership: A Sustainable Role for Europe*. Dordrecht: Kluwer.

Henikoff, J. (1997). 'Bridging the intellectual property debate: methods for facilitating technology transfer in environmental treaties', in L. E. Susskind, W. M. Moomaw, and T. L. Hill (eds), *Innovations in International Environmental Negotiation*. Cambridge, Mass.: Pon Books.

Héritier, A., (1996). 'The accommodation of diversity in European policy-making and its outcomes: regulatory policy as a patchwork'. *Journal of European Public Policy*. 3 (2): 149–76.

Héritier, A., Knill, C., and Mingers, S. (1996). *Ringing the Changes in Europe: Regulatory Competition and the Redefinition of the State, Britain, France, Germany*. Berlin: De Gruyter.

Hisschemöller, M. (1993). *De Democratie van Problemen, De Relatie Tussen de Inhoud van Beleidsproblemen en Methoden van Politieke Besluitvorming*. Amsterdam: VU uitgeverij.

Hisschemöller, M., and Gupta, J. (1999). 'Problem-solving through international environmental agreements: the issue of regime effectiveness'. *International Political Science Review*, 20 (2): 153–76.

IPCC (2001). *Summary for Policymakers: A Report of Working Group I of the Intergovernmental Panel on Climate Change*. Cambridge: Cambridge University Press.

Kandlikar, M., and Sagar, A. (1999). 'Climate change research and analysis in India: an integrated assessment of a North–South divide'. *Global Environmental Change*, 9: 119–38.

Lappe, M., and Bailey, B. (1999). *Against the Grain: The Genetic Transformation of Global Agriculture*. London: Earthscan.

Leow, W.-J. (1997). 'A trade cooperative for global forestry management', in L. E. Susskind, W. M. Moomaw, and T. L. Hill (eds), *Innovations in International Environmental Negotiation*. Cambridge, Mass.: Pon Books.

Maya, S., and Gupta, J. (eds) (1996). *Joint Implementation: Weighing the Odds in an Information Vacuum*. Zimbabwe: Southern Centre on Energy and Environment.

Noordwijk Declaration (1989). 'Noordwijk Declaration on Climate Change', in P. Vellinga, P. Kendall, and J. Gupta (eds), *Noordwijk Conference Report*, vol. i. The Hague: Ministry of Housing, Physical Planning, and Environment.

Nyerere, J. K. (1983). 'South–South options in Gauhar', in G. Altaf (ed.), *South–South Strategy*. London: Third World Foundation.

Oberthür, S., and Ott, H. E. (1999). *The Kyoto Protocol: International Climate Policy for the 21st Century*. Berlin: Springer.

Pearce, D. W., and Perrings, C. A. (1995). 'Biodiversity conservation and economic development: local and global dimensions', in C. A. Perrings, K.-G. Mäler, C. Folke, C. S. Holling, and B.-O. Jansson (eds), *Biodiversity Conservation*. Dordrecht: Kluwer.

Pearce, D., Cline, W. R., Achanta, A. N., Fankhauser, S., *et al.* (1995). 'The social costs of climate change', in J. Bruce, Hoesung Lee, and E. Haites (eds), *Climate Change 1995: Economic and Social Dimensions of Climate Change; Contribution of Working Group III to the Second Assessment Report of the Intergovernmental Panel on Climate Change*. Cambridge: Cambridge University Press.

Ramphal, S. S. (1983). 'South–South: parameters and pre-conditions', in A. Gauhar (ed.), *South–South Strategy*. London: Third World Foundation.

Roberts, A., and Kingsbury, B. (1993). 'Introduction: the UN's roles in international society since 1945', in A. Roberts and B. Kingsbury (eds), *United Nations, Divided World: The UN in International Relations*. Oxford: Clarendon Press.

Sands, P. (1990). *Lesson Learned in Global Environmental Governance*. New York: World Resources Institute.

Sands, P. (1992). 'The United Nations Framework Convention on Climate Change', *Review of European Community and International Environmental Law*. 1 (3): 270–7.

Schelling, T. C. (1960). *The Strategy of Conflict*. Cambridge, Mass.: Harvard University Press.

Schrijver, N. (1995). *Sovereignty over natural resources: balancing rights and duties in an interdependent world*. Groningen: Rijksuniversiteit Groningen.

Schrijver, N. (2001). 'Development—the neglected dimension in the international law of sustainable development', lecture at the Institute of Social Studies, The Hague, 11 Oct.

Sebenius, J. K. (1993). 'The Law of the Sea Conference: lessons for negotiations to control global warming', in G. Sjostedt (ed.), *International Environment Negotiations*. Laxenberg: IIASA.

Serageldin, I. (1999). 'Biotechnology and food security in the 21st century'. *Science*, 285: 387–9.

Shiva, V. (1993). *Monocultures of the Mind*. London: Zed Books.

South Centre (1993). 'An overview and summary of the Report of the South Commission', in South Centre (ed.), *Facing the Challenge: Responses to the Report of the South Commission*. London: Zed Books.

South Commission (1990). *The Challenge to the South: The Report of the South Commission*. Oxford: Oxford University Press.

SWCC Declaration (1990). *Ministerial Declaration of the Second World Climate Conference*. Geneva: World Meteorological Organization.

TERI (1998). *Climate Change: Post-Kyoto Perspectives from the South*. New Delhi: Tata Energy Research Institute.

Third World Network (1989). *Toxic Terror: Dumping of Hazardous Wastes in the Third World*. Penang: Third World Network.

UNDP (1996). *Human Development Report 1996*. Oxford: Oxford University Press.

UNEP (2000). *Global Environmental Outlook*. London: Earthscan.

8

Protecting the Vulnerable: Climate Change and Food Security

Thomas E. Downing

INTRODUCTION

IN the past several years debates about climate change have moved from speculation about likely impacts to how to detect and prepare for expected impacts. The Third Assessment Report of the Intergovernmental Panel on Climate Change (IPCC) is a major milestone. The nine major conclusions in the summary for policymakers (IPCC 2001: 3–9) are:

1. Recent regional climate changes, particularly temperature increases, have already affected many physical and biological systems.
2. There are preliminary indications that some human systems have been affected by recent increases in floods and droughts.
3. Natural systems are vulnerable to climate change, and some will be irreversibly damaged.
4. Many human systems are sensitive to climate change, and some are vulnerable.
5. The potential for large-scale and possibly irreversible impacts poses risks that have yet to be reliably quantified.
6. Projected changes in climatic extremes could have major consequences.
7. Adaptation is a necessary strategy at all scales to complement climate change mitigation efforts.
8. Those with the least resources have the least capacity to adapt and are the most vulnerable.
9. Adaptation, sustainable development, and enhancement of equity can be mutually reinforcing.

The conclusions from (4) onwards essentially concern social vulnerability and adaptive capacity: Who are the most vulnerable to climate change? What are the prospects for adaptation? How confident are we in relating present vulnerability to long-term threats of climate change? These questions concern vulnerable peoples, rather than ecosystems and places, and prospects for societal decision-making to reduce present vulnerability and long-run climate change impacts. This paper draws lessons for adapting to climate change from research and practice on food security.

The paper begins by setting the scene: What is the risk of climate change for food security? Who are the most vulnerable? Looking forward, the third section outlines the range of options for adapting to climate change and improving food security. The fourth section presents the notion of a vulnerability–adaptation science as a means to relate climate change to present vulnerability.

PRESENT VULNERABILITY AND FOOD SECURITY

Questions of Scale

Questions of identifying vulnerable peoples, the pressures to which they are vulnerable, and their capacity to adapt differ according to scale.

At the local scale, vulnerable populations and their livelihoods are often focused on households and relations with their communities (see e.g. Table 8.1). Vulnerability at this scale stems from the human ecology of production—the availability and use of resources, decision-making regarding land use and cultivation, gender and labour issues. Evaluations of detailed studies at this scale suggest that the sensitivity of individual crops needs to be minimized by agronomic adjustments (e.g. changed planting dates), adoption of other crops (e.g. ones adapted to shorter growing seasons), and new agronomic technologies (e.g. gene manipulation to enhance the CO_2 effects on water use efficiency to overcome precipitation changes).

At a national and regional scale, economic and policy institutions govern resources. Site-level adjustments can be supplemented with land use changes, such as shifts in the area devoted to specific crops. However, the main issues for food security at the national and regional scale are the operation of the market and the political economy of resource allocation.

At the global scale, the focus shifts to concerns for planetary food self-sufficiency and the distribution of agriculture between countries. For example, a recent update of global agro-ecological mapping provides a coherent analysis of climate change risks for agriculture (Fischer *et al.* 2001). Potential cereal production is likely to expand in large areas of northern mid-latitudes, especially in Canada, Russia, and China (at least

TABLE 8.1. *Sources of risk to household livelihood security*

Sources of livelihood	Types of risk				
	Environmental	Social		Economic	Conflict
		State	Community		
Human capital: labour power, education, health	Disease epidemics (malaria, cholera, dysentery) due to poor sanitary conditions, AIDS	Declining public health expenditures, user charges, declining education expenditures	Breakdown in community support of social services	Privatization of social services, reduction in labour opportunities	Conflict destroys social infrastructure, mobility restrictions
Financial and natural capital: productive resources (land, machinery, tools, animals, housing, trees, wells, etc.); liquid capital resources (jewellery, granaries, small animals, savings)	Drought, flooding, land degradation, pests, animal disease	Land confiscation, no secure tenure rights, taxes, employment policies	Appropriation and loss of common property resources, increased theft	Price shocks, rapid inflation, food shortages	Conflict leads to loss of land, assets, and theft

TABLE 8.1. (*cont.*)

Sources of livelihood	Types of risk				
	Environmental	Social		Economic	Conflict
		State	Community		
Social capital: claims, kinship networks, safety nets, common property	Recurring environmental shocks break down ability to reciprocate; morbidity and mortality affect social capital	Reduction in safety net support (school feeding, supplementary feeding, food for work, etc)	Breakdown of labour reciprocity, breakdown of sharing mechanisms, stricter loan requirements, lack of social cohesion	Shift to institutional forms of trust, stricter loan collateral requirements, migration for employment	Communities displaced by war, theft leads to breakdown in trust
Sources of income: productive activities, process and exchange activities, other sources of employment, seasonal migration	Seasonal climatic fluctuations affecting employment opportunities, drought, flooding, pests, animal disease, morbidity and mortality of income earners	Employment policies, declining subsidies or inputs, poor investment in infrastructure, taxes		Unemployment, falling real wages, price shocks	Marketing channels disrupted by war

Source: CARE, from Sphere workshop (Sphere Project 2001) .

for the Max Planck Institute for Meteorology ECHAM4 climate scenario for 2080, using present cropping systems including multiple rain-fed cereals). In contrast, tropical countries are more likely to suffer adverse consequences. The tally of winners and losers is shown in Figure 8.1. Gains in cereal potential are shown for twenty-four developed countries and sixty-eight developing countries, but the gains among developed countries are much the larger. Losses are modest for the seventeen developed countries, less than half of the losses shown for forty-nine developing countries.

Three conclusions regarding vulnerability and adaptation are illustrated by the Fischer *et al.* (2001) assessment. First, at the global level total food production is not a real threat. However, the distribution of food will require greater trade to balance out the winners and losers. Present cereal production among the winners is 568 million tonnes, against production in those countries that lose agro-ecological potential of 475 million tonnes. Among the winners, in the ECHAM4 scenario, potential cereal production increases by over 20 per cent (128 million tonnes), while the losers suffer losses of 7 per cent (33 million tonnes). Such impacts would require considerable increases in trade, particularly to cope with increases in demand related to population and economic growth in developing countries. Poor countries may not be able to afford additional imports unless international support is available.

Second, the regional impacts are quite diverse—and not predictable from the range of climate scenarios available to date. Effects will be patchy, often with regions of risk and benefit within countries. Highlands may gain in terms of potential cereal production while dessication and drought become worse problems in the semi-arid lowlands. Regional investment

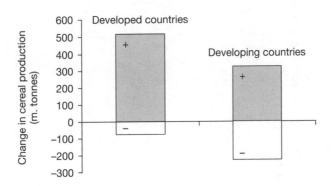

FIG. 8.1. Impact of climate change on cereal production in developed and developing countries, comparing 2080 to 2000.

Source: (Fischer *et al.* 2001).

TABLE 8.2. *Climate change impacts on cereal production potential in 2080 related to present prevalence of undernourishment*

Level of malnutrition	No. of countries	GDP (1995) ($ per capita)	% of population undernourished	Present cereal production (m. tonnes)	Cereal gap, (m. tonnes)	Climate impact on cereals (m. tonnes)
Losers						
Low	11	2,113	9	129	1	−12
Medium	19	486	23	307	9	−19
High	10	264	48	39	5	−2
Gainers						
Low	17	1,658	11	513	5	111
Medium	8	198	24	29	1	11
High	13	184	53	26	4	6

Source: Fischer et al. (2001).

may depend on the spatial patterns of expected changes. Agricultural capital investment is likely to be sensitive to medium-term prospects. Already planners need to identify regions where the return is likely to be sufficient for increased investment.

Third, at the local level, changes in livelihoods and food security are expected. Table 8.2 shows the agro-ecological impacts of climate change grouped according to losers and winners, each subdivided into three levels of existing malnutrition. The analysis is only indicative of the potential concurrence of climate change impacts on top of existing food gaps (the deficit between national production and requirements for a minimum diet). The ten countries with high rates of malnourishment might see their national food balance deficit increase, from 5 to 7 million tonnes per year. Perhaps equally worrying are the eleven countries with lower rates of malnutrition that might see their food gap increase by an order of magnitude (from 1 to 13 million tonnes). On the other hand, some thirty-eight countries would see gains in production that more than offset current food gaps. The focus on food security emphasizes the need for adaptation actions related to rural development options—such as health systems, employment (especially off-farm), and agro-technology.

Linking Present Vulnerability and Future Climate Change Impacts

The above agro-ecological assessment maps future climate change (say in the 2080s) onto present agricultural systems. A more dynamic approach would be to project present vulnerability and climate change, and the interactions that can be called impacts, on the same time scale. However, this is a difficult task, given the many social and economic futures that are possible on time scales of decades to centuries.

An alternative is to cluster communities, vulnerable populations, key sectors, or countries according to relative impacts and adaptive capacity. Figure 8.2 illustrates this approach, charting development opportunities and challenges according to present adaptive capacity and vulnerability to future climate change impacts. The four quadrants represent qualitatively different prospects. The most serious concerns are for the upper left quadrant, where present adaptive capacity is low and climate change impacts are likely to be high. For example, livelihoods in the coastal zone of Bangladesh are almost certain to fall in this category of vulnerable communities. Further refinement in climate predictions is not necessary to assign a high priority to increasing adaptive capacity. Where impacts could be high but adaptive capacity is also high—coastal resorts in Antigua, for example—development assistance could lead to new opportunities and a sustainable strategy to cope with increased risks (the upper right quadrant).

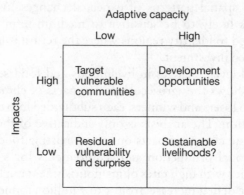

FIG. 8.2. Climate change adaptation matrix. The quadrants represent artificial boundaries of our knowledge of anticipated impacts of climate change and capacity of livelihoods or regions to adapt to climate change impacts.

Where impacts are low, further information on the nature of climate change may be warranted, although a precautionary approach to situations of low capacity may be suitable to be prepared for surprises (such as a rapid increase in damaging extreme events), as in the lower left quadrant. Where impacts are expected to be relatively low and capacity high, there should be little impediment to achieving sustainable livelihoods.

The schematic figure 8.2 suggests that some useful information can be gleaned about the potential impacts of climate change and adaptive capacity, at least to cluster situations in a relative sense. Here a first experiment, using existing data at the country scale, illustrates the approach.

The first step was to devise an index of aggregate climate change impacts. While there are an increasing number of global climate change impact studies, few include more than one sector at the country level. For illustrative purposes, an earlier study commissioned by the EU ExternE project (with support from the International Energy Agency's Greenhouse Gas R & D Programme) provides the requisite scale and coverage (Downing *et al.* 1994). Impacts are estimated by linking a climate change scenario (in this case only one—the IS92a linked to the early climate change scenario produced by the Goddard Institute for Space Studies) to a first-order biophysical impact model (such as a simple water balance). The change in the impact model is then valued according to various econometric techniques (none of which should be taken as wholly reliable).

The sectors included in the impact models at the country level are agriculture, biodiversity, water, cooling energy, heating energy, and sea-level rise. To compile an aggregate index, the country-level estimate of climate change damages for each sector were multiplied by a weighting factor (which values agriculture, water, and sea-level rise impacts higher than energy impacts), then the sectoral costs were added together and normalized. The quintiles of the impact index were determined. In the following clusters the first two impact quints (that is, 40 per cent of the countries) were considered high potential impacts.

The index for adaptation is much simpler: this experiment uses the Human Dimensions Index (HDI) (United Nations Development Programme 1999) as an indicator of relative adaptive capacity. The HDI is itself an aggregate of indicators on life expectancy, Gross Domestic Product (GDP) per capita, and education.

The countries were grouped according to their estimated impacts and adaptive capacity, using the 2×2 matrix shown in Figure 8.2.

The most vulnerable countries (high impacts and low adaptive capacity) contain over 3 billion people, with an average per capita GDP of $2,200 (Fig. 8.3). The opposite situation—where sustainable livelihoods are not likely to be threatened by climate change (low impacts, high capacity) contain less than 0.5 billion people, and their average GDP is nearly $12,000.

Many wealthy countries can also expect high climate change impacts (in terms of the global cost of climate change); nearly 1.5 billion people live in these countries. The most uncertain class is where the aggregate index of impacts is relatively low, but adaptive capacity is also low; in these countries of residual risk climate change could contain some surprise or be a serious threat to some livelihoods within those countries. They comprise about the same number of people as the sustainable livelihoods cluster, but with about one-fourth of the GDP per capita.

| | Adaptive capacity | | | | Adaptive capacity | |
	Low	High			Low	High
High	3,247	1,349		High	2,237	15,434
Low	581	424		Low	3,289	11,777

Impacts (left matrix) — Population (m.) (total: 5,602)

Impacts (right matrix) — GDP per capita ($) (average: $6,245)

FIG. 8.3. A preliminary estimate of adaptive capacity and vulnerability to climate change impacts. Population and GDP per capita are for 1995. Totals and average at the bottom are for the 162 countries included in the data set.

PROSPECTS FOR ADAPTATION

Range of Adaptation Measures for Climate Change and Food Security

The range of adaptive responses for agriculture and food security is shown in Table 8.3. The measures and strategies are roughly ordered according to the scale of implementation, from local farm management and livelihood security to national and international action. Many of the actions can be called 'no regrets', that is they have widespread benefits at present. However, some are costly and require significant infrastructure and development funding. Other measures can lead to capacity-building, or require a strategic framework to be implemented.

TABLE 8.3. *Adaptive responses to climate change for food security*

Measure or strategy	No regret	Infrastructure	Strategic capacity-building
Reduce production subsidies	✓		
Drought preparedness, warning, and management	✓		✓
Avoid monoculture, diversify crops	✓		
Conserve soil moisture and nutrients	✓		
Diversity income, off-farm employment	✓		✓
Crop, farm and income insurance			✓
More R&D on heat and drought resistant varieties; maximize CO_2 effects			✓
Tailor land use planning to consider potential climate change			✓
Reduce run-off, improve water uptake, reduce wind erosion	✓	✓	
Increase irrigation efficiency; prevent salinization	✓	✓	
Upgrade food storage and distribution systems	✓	✓	
Liberalize agricultural trade	✓		✓
Increase food aid			✓
Economic safety nets, welfare systems			✓
Change dietary preferences			✓

Stakeholders in Adaptation

Adapting to climate change is not automatic. The motivations, constraints, and domains of authority of decision makers involved in shaping policy, implementing decisions, and coping with the consequences of changes in resources and hazards must be considered (see Table 8.4). The principal stakeholders range from vulnerable consumers to international organizations charged with research and relief (Downing *et al.* 1997). Stakeholders will suffer the consequences of climate change to varying degrees and have different roles for different types of adaptations. This is likely to influence their involvement in planning and implementing adaptive responses.

Consumers of both food and water are the ultimate stakeholders in adapting to climate change. For particularly vulnerable groups (such as resource-poor farmers, landless labourers, urban poor, the destitute and displaced or refugee populations), the outcome of strategies to adapt to climate change and climatic hazards may alter their livelihoods.

Producers also have varying interests in climate change. Subsistence farmers are less likely to have the resources to consider anticipatory action than large-scale commercial farmers. Commercial farmers are more likely to be linked to national markets and international agribusinesses and to be able to invest in agricultural technology.

One of the key stakeholders in enacting forward-looking strategies is business, ranging from local market traders to international commodity and research organizations. However, commodity traders are not likely to be affected directly by the consequences of climate change, provided that production is viable and trade is required somewhere in the world. Incentives may be required to induce agribusiness to adopt longer planning horizons and to develop and implement adaptive responses.

The bulk of responsibility at present for designing, evaluating, and implementing strategic responses (anticipatory actions, planning institutional change, and research and education) rests with national governments, national and international research centres, and aid organizations (particularly bilateral and multilateral, although some international non-governmental organizations take an interest in adaptation policies).

Strategic Adaptive Responses

Four groups of policies offer the kinds of opportunities that could form a strategic plan for agriculture and food security. Evaluating such strategic plans requires screening against a range of criteria (Grimble and Chan 1995), including impact on stakeholders, resilience and effectiveness,

TABLE 8.4. *Stakeholders and adaptive responses*

Stakeholders	Adaptive responses				
	Consequences	Anticipatory	Institutional and regulatory	Research and education	Development and assistance
Vulnerable consumers	√				+
Subsistence producers; private water carriers	√				+
Commercial producers		+		+	√
Market traders	+				√
Irrigation, water, and sewage boards		?	?	+	√
Food processing and trading; river basin development agencies		?	?	√	√
National and international research				+	+
Government ministries (planning, agriculture, health, and water)		+	+	+	√
Aid and community development organizations		+	?	+	√

Notes: 'Consequences' refers to bearing the consequences of climate change impacts, that is those stakeholders who are directly affected by altered agricultural production. The adaptive responses correspond to the guidelines: 'Anticipatory' represents anticipatory adaptation and strategies targeted for coping with climate change; 'Institutional and regulatory' denotes institutional and regulatory adaptation to prevent increased vulnerability; 'Research and education' represents research and education to develop and implement new solutions; 'Development assistance' represents development assistance that implements current options for sustainable agricultural development and reducing vulnerability to climatic hazards.

The ratings indicate the type of response likely to be of interest to each stakeholder: √ indicates primary interest in adapting to climate change; + indicates secondary, but important, interest in adaptive strategies; ? indicates uncertain but potential role in adaptation.

strategic importance, timing of benefits, economic evaluation, and specific constraints and barriers.

First, common *farm-level adaptive responses* include substitution of agronomic practices, altered inputs, and agricultural development. The priority stakeholder is likely to be the smallholder farming sector. Commercial farms would be less likely to need assistance in these sorts of adaptive strategies. On the other hand, these strategies are less likely to be effective for agro-pastoralists and pastoralism in general. There may be some competition for development assistance, but in general these agronomic packages do not entail serious resource use conflicts between farming groups and others or within farming communities.

Given modest resources and political will, it should be possible to implement all of the adjustments relatively quickly, often within a single season. Even investment in soil and water conservation represents a relatively modest investment, compared with the potential risk of climate change. Agronomic improvements are effective, can be readily implemented, and have few substantial constraints to their adoption.

Irrigation schemes may be a special case in considering responses to climate change. They are sensitive both to direct impacts of climate change and to changes in water supply. The design and investment cycle is such that major schemes are expected to be operational for at least several decades, which puts irrigation planning into the time scale of expected climate change.

Second, *national economic policies* can maintain a positive food balance and exports. Maintaining strategic reserves allows government (or marketing bodies) to dampen price fluctuations and release food in emerging crises. Quite large national reserves have been held in the past few decades, in some countries enough food to meet consumption for a year or more. In the 1990s these reserves were reduced under structural adjustment agreements. International lending organizations noted that such reserves are costly to maintain and absorb a significant fraction of government resources. Better monitoring and more timely responses were seen as more efficient ways to cope with food shortages.

An alternative strategy would be to adjust markets and trading conditions to promote private sector responses to climate change and climatic variability. This might take the form of tax incentives for carry-over stocks or bonds to smooth income between adverse and good trading years. The stakeholders in such a strategy would be producers, including market traders, millers, and agribusiness in general. This kind of strategy would build upon present efforts towards reduced trade barriers, with some specific adjustments to accommodate to climate change.

National or regional planning to promote general agricultural development does not require additional action specifically because of climate change. However, the gap between research (as in potential yields) and practice (the present yields achieved by marginal farmers) is large in Africa. The need to adapt to climate change could be used as one argument for fresh initiatives in promoting adaptive agricultural research and development, especially in Africa.

The primary beneficiaries of national economic planning are consumers and commercial producers, who depend on markets for food consumption. Market adjustments may entail some trade-offs between consumers and producers, or between relatively prosperous farmers and vulnerable smallholders, who may not have access to inputs and markets. Yet, the potential for multiple benefits is high (except for strategic reserves, which are a burden on the economy). Taken as a group, these strategies would be reasonably effective in preparing for climate change.

These strategies can be readily implemented, are not likely to have irreversible impacts (depending on the nature of specific developments), and generally have a strategic role in promoting a resilient economy. Benefits could be realized throughout the economy, although investments would take time to achieve demonstrable results.

The third group of policies is at the *global level*, concerned with investment, demand, and trade. As for national and regional economic development, suggested strategies range from building strategic reserves to encouraging free trade and transfer of agricultural technology.

The arguments for building global strategic reserves, both for major foods and of financing, follows the same logic as at the national level, with the additional notion of international responsibilities to share the burden of climate change impacts. Additional trade would smooth out fluctuations in national production. Maintaining international prices within acceptable limits would benefit poorer countries, which might not be able to afford large imports in times of scarcity. It should be more effective to hold reserves at the global level, or shared among regional trading partners, than for each country to buffer its internal production. In the transition towards a new climate, such an international capacity to prevent food deficits from becoming survival emergencies appeals to humanitarian goals of ending famine and reducing hunger.

Encouraging free trade between countries should stimulate agricultural markets in regions with a comparative advantage. This may be a major benefit to some countries, and a significant cost to others, as the impacts of climate change alter traditional markets. In principle, free trade allows national surpluses and deficits to be accommodated more efficiently. Supply and price fluctuations are thus buffered at the global level, widening the potential pool of responses to climate change. Free trade, of course, is already on the international agenda and little further encouragement is required. However, some incentives to the private sector to absorb additional risks may be required.

The most costly and long-term strategy proposed in this review is to develop international mechanisms to promote agro-technology transfers to developing countries. An initial agreement might focus on basic foodstuffs: wheat, rice, and maize. International agencies might license new technologies developed by biotechnology firms for dissemination and use in developing countries. Adaptation and abatement might be explicitly linked, with a requirement that beneficiaries have agreed to limit greenhouse gas emissions. Funding requirements would be significant, although connections to emission taxes could be made.

The immediate beneficiaries of global linkages are commodity brokers and private companies, although aid and government agencies have strong interests. As for national policies, some resource allocation issues may imply trade-offs between regions, commodities, and farming populations. Global policies should be highly effective, with a relatively low specificity to climate change or dependence on specific climate scenarios. The planning horizon is of the order of five years, with benefits realized somewhat later. Except for agro-technology transfer, there are few constraints to implementing global policies (other than bureaucratic inertia and funding).

Finally, *drought hazard and vulnerability* are present risks and are likely to be the most damaging locus of impacts of climate change. Concerted action is required to reduce vulnerability by monitoring both drought and vulnerability and by preparing to respond effectively to emerging crises. Considerable progress has been made in the past decade; a further decade of development might reap substantial rewards in efforts to eliminate widespread famine and enhance livelihood security, at least in times of drought. An example of matching scales of risk to interventions for famine early warning systems is shown in Table 8.5.

The priority stakeholders should be the most vulnerable socioeconomic groups affected by drought crises, although many levels of local, national, and international actors are required to implement drought monitoring and mitigation, and emergency responses. Drought policies provide multiple benefits to the extent that they contribute to development in general by reducing investment risk. However, crisis interventions can be counter-productive if they create a dependency syndrome or compete for resources from other activities with longer-term benefits.

Reducing risk can have strategic importance. Increased monitoring and response capabilities should improve the ability to respond to long-term climate change and to economic management of the agricultural sector in general. With foresight, crises can be used to promote sound resource policies, although this remains the anomaly in most of the world rather than

TABLE 8.5. *US Famine Early Warning System vulnerability matrix*

Level	Conditions	Interventions	Surveillance
Famine	Destitute: coping strategies exhausted	Emergency relief: food, shelter, medicine, etc.	Monitoring access to relief
Extreme or at risk	Altered production: liquidating productive resources (e.g. planting seed, tools, oxen, land, breeding animals, whole herds) or abandoning preferred means of production in favour of emergency (non-traditional) sources of income, employment, or production (e.g. migration or workers or whole families)	Relief and/or mitigation: support nutrition (e.g. food relief), income (seeds), and assets (fodder)	Intensive community, household, and/or individual assessment through on-site surveys to measure specific needs and resources and identify appropriate responses
High	Depleting assets: liquidating wealth but not yet the means of production, for example by selling livestock, bicycles, and possessions. Disrupting production: coping strategies have significant individual, household and/or environmental costs (e.g. increased wage labour, gathering of wild food, selling firewood, farming marginal land, labour migration, borrowing at high interest rates)	Mitigation and/or relief: support income and assets (food- and cash-for-work, etc.)	Rapid area and population appraisals: increase information about specific areas and vulnerable groups through site visits and other means

Moderate	Drawing down assets: liquidating less important assets, husbanding resources, and minimizing expenditures through, e.g. drawing down food stores, reducing food consumption, sale of surplus small stock. Maintaining production strategies: only minor stress-related changes in income strategies (e.g. altered cropping practices, increased remittances from relatives)	Mitigation and/or development: support assets (release food stocks, stabilize prices, subsidize fodder, open community grain banks, etc.)	Targeted area and population monitoring: increase information on specific issues and vulnerable groups, but without mobilizing substantial new resources
Slight	Maintaining or accumulating assets: preferred production strategies are used to cope with seasonal stress and to maintain or increase wealth	Development: long-term strategies to reduce vulnerability	Regular collection of information, primarily for development planning

Source: Based on a US Famine Early Warning System (FEWS) internal memorandum, 19 Apr. 1991.

the norm. Once drought monitoring, mitigation, and preparedness are operational, the benefits are recurrent, although they are only significant during times of potential crisis. There are few irreversible impacts to monitoring and preparedness and the initial investment can be fairly modest.

Constraints include the need for sustained information collection, processing, and reporting, often requiring significant development of technologies and organizations. While most planners and vulnerable populations agree that drought hazard planning and reduction are desirable, the lack of reliable systems in most countries implies further social, economic, and political constraints. One that is commonly cited is the short attention cycle: drought planning typically peaks about a year after the drought and is forgotten until the next crisis. If drought becomes more frequent with climate change, this deficiency may be overcome.

METHODOLOGY FOR A VULNERABILITY–ADAPTATION SCIENCE

The conclusions presented above rest upon a decade and more of climate change impact assessment and vulnerability research. Here I comment on the methodology that will deliver a robust science.

Most research in the 1980s and 1990s on climate change impacts adopted one of two approaches. Regional (and some global) studies tended to be sectoral assessments, often linking climate scenarios with biophysical impacts or studying responses of individual 'exposure units' to selected climatic threats (and less often opportunities). Global integrated assessments (and also a few regional studies) that looked across sectors tended to do so based on abstract mechanisms (e.g. markets to mediate resource allocation) or with limited and constraining interactions between sectors. For example, one integrated assessment model predicted continental deforestation in Africa because land was required for achieving food self-sufficiency, even though Africa currently imports a large fraction of its food supply. Few (if any significant) studies related the dynamism of local vulnerable groups to processes and shocks at regional to global scales.

In contrast, research and practice on food security and famine early warning in this same period has sought more sophisticated means to integrate the local and global, to understand exposure across a range of shocks, and to decipher the climate–food chain of causation against a background of the human ecology of production, exchange entitlements, and political economy (Bohle *et al.* 1994).

Research on food security and famine early warning provide a foundation for a useful science of vulnerability and adaptation to climate change:

Conceptualizing vulnerability. The early models of food security adopted the approach of gathering all possible (or measurable) indicators, adding them up and deducing vulnerability. This was termed 'hoovering': vulnerability was the weight of the bag after vacuuming up everything in sight. The danger in the indiscriminate use of indicators is that many are correlated with each other, have weak or delayed connections to emerging vulnerability, and may miss indicators that are difficult to measure. Subsequently analysts emphasized logical, conceptual approaches to vulnerability, and used conceptual models to choose relevant indicators. Many agencies appear to have converged around a broad framing of vulnerability (as in the CARE approach, Fig. 8.4). Thus, the first step is to conceptualize the relevant dimensions of vulnerability and adaptation.

Social vulnerability. A critical concept is that vulnerability is characteristically about people rather than places. That is, vulnerability is a social phenomenon relevant to particular social groupings, whether demographic (elderly, young), economic (livelihoods, entitlements), or political (marginalized). It is people who are exposed (ultimately) to climate change. Nature–society interactions are at the heart of vulnerability assessment.

Integration. Concepts of vulnerability should integrate exposure at the level of vulnerable groups; therefore vulnerability must span across sectors and sources of shocks.

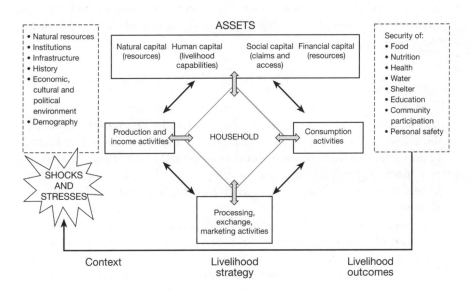

FIG. 8.4. CARE's livelihood security framework.

Source: Sphere Project (2001).

Relative measures. Vulnerability is not an external characteristic that can be universally described and observed. Rather it is a way of addressing complex problems and deciding a course of action—whether mobilizing aid to prevent famine or mapping underlying food insecurity to target food aid programmes. The analyst must choose indicators, aggregation models, and relevant thresholds for action.

Targeting and specificity. The clear need to target the most vulnerable groups and to identify risks in specific situations has led to increasingly sophisticated assessments. For example, early assessments that focused on food deficits may have missed the impacts of economic shocks.

Dynamic pathways and multi-level processes. Vulnerability changes quickly, the product of changes at several scales. Assessments focus on current situations, but in the context of time series and analogues to previous crises.

Response orientation. Vulnerability assessment grew out of a need to trigger relief, and still keeps a close connection with at least an implied decision-making framework (Fig. 8.4). Agencies implementing vulnerability assessments seek to distinguish between their own mandate and broader issues, and to collect appropriate information at least for their own decision-making.

Standards. Recently disaster relief agencies have worked to create and implement a series of standards, including standards and suggestions revisions for food aid. While not universally accepted, the process and experience around the world is proving helpful to discussion of ways to promote effective disaster aid.

CONCLUSION

Food security assessment and famine early warning are mature endeavours showing considerable progress over the past decade. This is itself a measure of increasing adaptive capacity with respect to climate change. The adaptive measures needed to promote food security in the face of climate change are known. But will the measures be adopted? Will they benefit the most vulnerable—the 3 billion people at risk of high impacts and with low adaptive capacity?

Almost ten years ago the Oxford Conference on Climate Change and World Food Security (Downing 1996) posed the question: How significant is the risk of climate change for regional collapse? The workshop concluded that climate change over the next decades was unlikely to seriously affect global food production. We also believed that individual livelihood security would come under increasing stress in some locales, although those might not be easily identified. We were not confident in our ability to analyse the risks at the intermediate scale of regions, and we are still not able to answer this fundamental challenge.

Much depends on global leadership. A 'new world order' might reinvigorate the development agenda. Or, on a more pragmatic level, adaptation funds in the climate change convention could embed management of climate change risks in present sustainable development planning.

Caution is necessary. I noted above that changes in agro-ecological potential would lead to an increase in world food trade. Under some scenarios of emission pricing, poor countries and consumers may not be able to command food at world prices that include high transport costs. Within regions and nations the engine of food security for rural households is investment in agriculture. This may be problematic in a world where more crops can be grown more cheaply in higher latitudes and where climate forecasts suggest increased risks (as in semi-arid tropical regions). Although livelihood security is a 'good thing', even the present modest international development targets are unlikely to be achieved (High-Level Panel on Financing for Development 2001). Linking global change and poverty alleviation makes sense, but there is no guarantee that the synergies are sufficient to redress fundamental disparities in world economic and political systems.

The research task to link sustainable livelihoods to global change forms the basis for what may be termed 'vulnerability–adaptation science' (see National Research Council 1999 for the broader perspective of sustainability science). Vulnerability–adaptation science seeks to extend the lessons learned over the past decade in food systems research. A particular concern is to link climate change policy to research and action on sustainable development and disaster mitigation.[1] Coping with existing climatic risks is a first step, and provides insight into mechanisms for adapting to climate change.

REFERENCES

Bohle, H. G., Downing, T. E., and Watts, M. J. (1994). 'Climate change and social vulnerability: towards a sociology and geography of food insecurity'. *Global Environmental Change*, 4 (1): 37–48.

Downing, T. E. (ed.) (1996). *Climate Change and World Food Security*. Berlin: Springer.

[1] The International Geographical Union Task Force on Vulnerability, coordinated by the Stockholm Environment Institute, Oxford, and the International Network on Climate Adaptation, coordinated by the Potsdam Institute for Climate Impacts Research, are international networks that seek to develop this research and policy agenda.

Downing, T. E., Greener, R. A., and Eyre, N. (1994). *Global Emissions and Impacts: Report to the International Energy Agency*. Oxford: Environmental Change Unit.

Downing, T. E., Ringius, L., Hulme, M., and Waughray, D. (1997). 'Adapting to climate change in Africa: prospects and guidelines'. *Mitigation and Adaptation Strategies for Global Change*, 2: 19–44.

Downing, T. E., Harrison, P. A., Butterfield, R. E., and Lonsdale, K. (eds) (2000). *Climate Change, Climatic Variability and Agriculture in Europe: An Integrated Assessment. Final Report of the CLIVARA Project*. Oxford: Environmental Change Institute.

Fischer, G. S., van Velthuizen, H. M., Nachtergaele, F. O. (2001). *Global Agro-Ecological Assessment for Agriculture in the 21st Century*. Laxenburg: IIASA and FAO.

Grimble, R., and Chan, M. K. (1995). 'Stakeholder analysis for natural resource management in developing countries: some practical guidelines for making management more participatory and effective'. *Natural Resources Forum*, 19 (2): 113–24.

High-Level Panel on Financing for Development (2001). *Recommendations of the High-Level Panel on Financing for Development*. New York: UN.

IPCC (2001). *Summary for Policy Makers: Working Group II: Climate Change 2001: Impacts and Adaptation*. Cambridge: Cambridge University Press.

National Research Council (1999). *Global Environmental Change: Research Pathways for the Next Decade*. Washington, DC: National Academy Press.

Sphere Project (2001). *Report of an Inter-Agency Workshop to discuss Minimum Standards for Food Security in Disaster Response, Oxford, 2–3 July 2001*. Oxford: Oxfam.

United Nations Development Programme (1999). *Human Development Index*. New York: UNDP.

INDEX